The Open University

T356
Engineering small worlds:
micro and nano technologies

BLOCK 3

WORKING WITH NATURE: THE CELL AS AN ENGINEERING SYSTEM

The Open University
Walton Hall, Milton Keynes
MK7 6AA

First published 2006.

Edited and designed by The Open University.

Typeset by The Open University

Printed and bound in the United Kingdom by Thanet Press Ltd, Margate.

ISBN 0 7492 0253 X

1.1

This publication forms part of an Open University course T356 *Engineering small worlds: micro and nano technologies*. Details of this and other Open University courses can be obtained from the Student Registration and Enquiry Service, The Open University, PO Box 197, Milton Keynes, MK7 6BJ, United Kingdom: tel. +44 (0)870 333 4340, email general-enquiries@open.ac.uk

Alternatively, you may visit the Open University website at http://www.open.ac.uk where you can learn more about the wide range of courses and packs offered at all levels by The Open University.

To purchase a selection of Open University course materials visit http://www.ouw.co.uk, or contact Open University Worldwide, Michael Young Building, Walton Hall, Milton Keynes MK7 6AA, United Kingdom for a brochure. tel. +44 (0)1908 858785; fax +44 (0)1908 858787; email ouwenq@open.ac.uk

CONTENTS

BLOCK 3 OVERVIEW

In this block, as we venture further downwards into the realm of the nanoscale, we inevitably find that much of what we wish to study is underpinned by basic principles of chemistry.

Furthermore, it soon becomes apparent that many of the ultimate goals of the nanotechnologist – smart materials that can repair themselves, sensors capable of detecting single photons, phonons or molecules, high-efficiency solar converters, motors smaller than a grain of pollen, self-replicating molecular machines and many more – already exist within living cells. So there is also much for the nanotechnologist to learn from biology.

The aims of this block are to introduce you to just enough chemistry and biology to begin to understand the basic physical principles on which nature's nanomachines operate, to show a few examples of what they can do, and to demonstrate some of the ways in which we are already putting this knowledge to work in practical engineering devices.

1 INTRODUCTION

1.1 Why study cells?

You may be surprised to find a block focusing on living cells in a course about engineering, but the Introductory study block will have given you a taste of some of the important new developments that are emerging in this area. In this block you have the chance to find out more. Indeed, many of the most exciting advances in nanotechnology are happening at the interface between engineering and biology.

So, are you thinking, 'Why should I be studying this? What's it to do with engineering?' In a few years' time, perhaps, the answers will be obvious. A parallel might be an engineer asking early last century, 'Why should I study the behaviour of electrons? What's it to do with engineering?'

EXERCISE 3.1

We already use nature's cells in large-scale industrial processes. Name one such process based on the use of cells.

If you can't wait a few years, here are some reasons why bionanotechnology is an important and rapidly developing branch of nanotechnology.

- The computing power of the human brain is unmatched by any electronic super-computer in power and compactness, and without the need for liquid helium or nitrogen cooling systems.
- The reaction extrusion process that produces spider silk is a process and product unrivalled by anything in industry.
- The ways in which your arm, for example, can detect the position of and forces on different parts of it, use feedback to apply precisely graded forces to do things and detect and repair damage to itself again are far superior to any smart materials and structures that we are capable of producing.
- Sensors in the eye, ear, nose, tongue and skin can detect signals with sensitivities that we can rarely match.
- Branches of trees develop tension wood and compression wood in response to, and to sustain, the gravitational forces on them: another natural smart material.

Examples like these (and there are many, many more) should convince even the most sceptical that nature can do engineering in ways and at scales that we still can't mimic. We need both to learn how to imitate nature's products synthetically and, where we lack the techniques or the understanding to imitate nature successfully, to develop increasingly sophisticated ways of harnessing nature to make things for us.

In any case, our views of what constitutes engineering are broadening. For instance, it's arguable that the biochemists who develop and manufacture pharmaceutical drugs, or the chemists who manipulate molecules to construct new polymers, are as much engineers as those who produce motor cars.

Nature provides us with a stunning array of highly sophisticated nanoscale devices and machines. A great deal has been found out about their underlying structure using established techniques such as electron microscopy and X-ray diffraction. As new instruments, such as the atomic force microscope you studied in Block 1, allow scientists to probe their properties on ever smaller scales (see, for example, the AFM image of DNA in Block 1 main text Figure 1.2), our understanding of how nature works is improving to the point where we can begin to apply or adapt the same principles to design useful devices of our own.

The fundamental unit of living systems is the cell, so I've decided to take the cell as the basis for this text. A cell is the simplest unit capable of independent existence. There are many different types, from the simplest single-celled bacteria to the complex and varied cells that make up plant and animal tissue. They range in size from about a micrometre to a millimetre, though the components within them are much smaller; nature has been 'working' at the nanoscale for a very long time. These cells can perform a huge range of functions, but the molecular components of all cells are remarkably similar.

Figure 3.1(a) shows a schematic diagram, such as you might find in a biology textbook, of the 'typical' features of animal and plant cells. You will learn more about many of the structures illustrated as you work through this text. In the cartoon in Figure 3.1(b) I've attempted to represent the functions of the different components in a more familiar way, to show how the cell might be considered as an 'engineering system' – although these shouldn't be taken too literally!

As this text is structured around the workings of a cell, it will inevitably involve touching upon biological ideas and language, and some of this is likely to be new to you. However, the main focus here is not the cells themselves, but the nanoscale constructions that exist within them. We'll look at the structure and function of a selection of these, and see some of the ways that we may be able to adapt them in the engineering of the future. The background science you need to understand this will be introduced as required.

A glossary of the biological terms used in Block 3 is given at the end of the *Principles* wing.

Our treatment will inevitably be very selective, but after studying this block you should feel more confident in interpreting the language of biology, in recognizing the parallels and the differences between these very small-scale systems and those you have studied before, and in appreciating the huge potential that this new collaboration between biology and engineering can offer. By the end of the block you'll be able to view a cell as something akin to a tiny industrial complex, with manufacturing plant, distribution systems, communication networks, power plants, and a whole host of associated machinery to keep everything working smoothly, safely and effectively.

1.2 Opportunities and constraints

Let's consider the range of functions that cells need to perform; this will give us a good idea of the sort of nanomachinery that might be available through working with nature. I've said that a cell is the fundamental unit of living systems, but what does that imply? Setting aside several centuries of philosophical debate about what defines a living organism, I'm going to accept this basic specification: the ability to reproduce, and to respond to and extract sustenance from the environment. The term 'organism' used here could be anything from a single cell to a complex collection of different cells fulfilling specialized functions – like you.

The first of these requirements – reproduction – is the one that most sets an organism apart from the other systems you have studied. Reproduction requires cells to divide, so they need a mechanism for creating copies of themselves. In most cases we want these copies to be faithful reproductions of the original, but a small amount of random change can also provide the means for improvement through evolution. You're probably already familiar with DNA as the genetic material that carries information

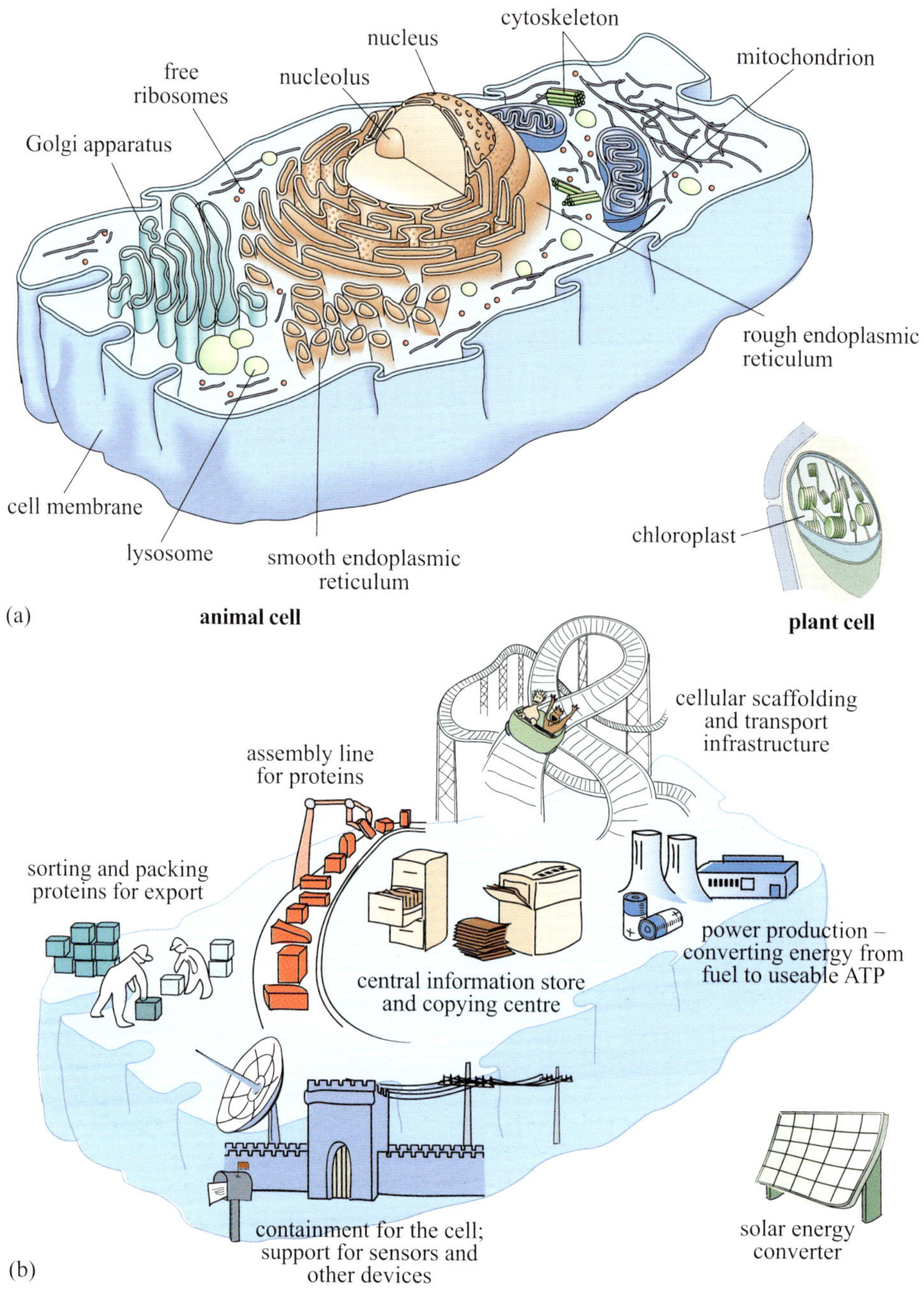

Figure 3.1 Schematic diagram of a cell as it might be viewed by (a) a biologist and (b) an engineer. The main structures found in cells are illustrated, and their primary functions indicated

about cells and lies at the heart of an almost perfect copying mechanism. In Figure 3.1(b) this is represented by a filing cabinet and photocopier: we'll investigate how it really works, and how the information it carries is used to assemble new material, in Section 6 Manufacture.

Organisms need to respond to their environment. This statement covers a multitude of possibilities, from slight adjustments arising from, say, a change in the weather, to large-scale mobilization of forces against attack from hostile invaders. Individual cells are constantly receiving and responding to signals, both from the surrounding environment and from their near neighbours. This is represented by a satellite dish and a postbox, and by cables connected to the cell in Figure 3.1(b), but the actual methods for receiving and passing on signals in cells are rather different from those we are used to. These will be considered in Section 3 Communication.

Organisms must be able to extract sustenance – chemicals and energy – from their environment. All machines that perform useful work require a source of energy, and cells are no exception. Energy is needed for the creation of new material from molecular building blocks. It's also required to power the many processes that go on in living things: processes such as nerve activity, cell division, muscular movement, and the transport of materials from one place to another. We also need to keep warm, but not too warm! The way that cells use energy has to be rather different from, say, burning petrol in a car, because of the limited temperature range over which living things operate. The generation of energy in a usable form and the methods for storing it (the power station and the batteries from Figure 3.1(b)) are the subject of Section 4 Energy requirements. The way that energy is used by cells to generate movement is examined in Section 5 Transportation.

Also implicit in this discussion is the need for a cell to maintain its integrity. Some sort of containment is needed to define the boundaries of the cell, to protect it and the processes going on within it, while allowing it to interact with other cells and the external environment. This will be my starting point in Section 2 Structural materials, where I consider the materials that give structure to the cell.

There is obviously much we can learn about building nanomachinery from studying cells, but it's also worth noting a few constraints. Think of the cells that make up your own body: what conditions do they need to work? When do they stop working properly?

EXERCISE 3.2

Suggest some of the conditions required for the nanomachines that operate within the human body.

The machinery within cells is optimized to work effectively over quite a narrow range of temperature and pressure (dependent to some extent on the type of cell). Most of it works only when surrounded by water, and the solution mustn't be too acidic or too alkaline, i.e. the pH must stay constant. A constant supply of energy and materials is required to keep things working. Change any of these conditions too much and we die. In fact, we die anyway, so there is obviously some degree of planned obsolescence in the system. The cell is a dynamic environment, with many bionanomachines created when required and then demolished, often lasting only a few seconds. Consequently their structures must be easy both to build up and to break down, and the energy changes required are small. The energy supply must therefore be both constant and in appropriately small packages for the cell's use.

These conditions suit the cell well, but they may not all be appropriate for applications outside cells. So we need to develop methods not only for copying designs from nature, but also for modifying them and adapting them to fit our needs. Unhindered by the legacy of evolution, which has delivered systems that work but not necessarily in the most effective or efficient way, we might even hope to find completely new ways of using biological materials.

1.3 BioMEMS

The development of micro and nano scale devices that incorporate or interact with biological systems is a new and rapidly advancing field. During our journey through the cell I'll include some examples of where the lessons learned from nature are starting to be applied in other contexts. The term bioMEMS (biomedical or biological microelectromechanical systems) is often used as a general term for devices or systems developed in this area, whether or not they actually include any electromechanical components. The term encompasses applications of biology to micro and nano scale systems and materials, as well as applications of micro and nano scale systems to biological and biomedical problems. The different types of application that result are summarized in Figure 3.2.

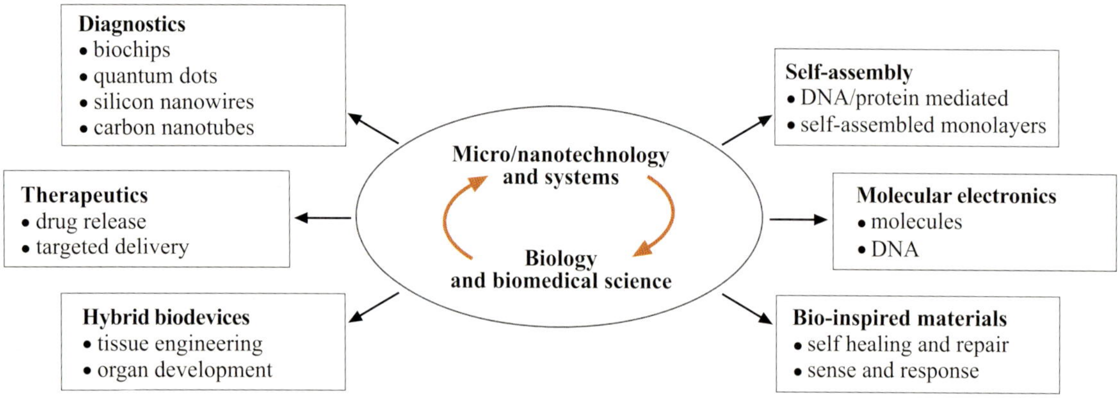

Figure 3.2 Research areas resulting from the integration of micro and nano scale systems and biomedical sciences

Many of the current applications of bioMEMS are medical. Devices that are to be put into the body, such as vessels for drug delivery or scaffolds for tissue repair, tend to follow similar design principles to those found in nature. If the devices are to be used externally, for instance for the detection of toxins or to look for particular sequences of DNA, then there is more scope for adapting biological design features to produce new types of structure. The terms *in vivo* (within a living organism) and *in vitro* (in an artificial environment outside the living organism) are often used to distinguish between these two situations.

While some applications are closely based on designs found in nature and others adapt these principles to different external environments, you may also find examples of synthetic molecules designed as alternatives to their biological counterparts, to fulfil a specific function, or of biological materials used in ways not found in nature.

Some areas of bioMEMS are already well developed; for example, lab-on-a-chip miniaturized analysis systems are widely available. Others have yet to find practical applications. However, as research continues and our understanding improves, it can only be a matter of time before more are incorporated into the engineering systems of the future.

1.4 Soft, wet and sticky

Before embarking on our tour of the cell, it is useful to set out a few general principles. When trying to pin down the fundamental differences between the tiny devices you'll study here and those in the preceding blocks, the three words that seem best to sum it up are soft, wet and sticky.

Soft, because instead of the silicon and metals that have provided the major raw materials until now, nature's machines are based on polymers and small molecules. Polymer chains are, to varying extents, flexible; they can fold up into complicated shapes, and change their shape in response to external forces. Polymer behaviour is often more complex than that of metals and ceramics, because different parts of the chain interact in different ways. The main carbon chain (that forms the backbone of the polymer) and the side groups attached to it are linked together by strong covalent bonds that may be able to rotate, but are very hard to break. However, weaker secondary bonds can also form between different parts of the chain, or between one chain and another, and these give a secondary structure to the chain that is often crucial in determining its properties. The resilience of nylon, for instance, is due not only to the strong bonds that hold the chain together but also to weaker bonds between parallel chains that make it harder for them to slide past each other; see Figure 3.3. Because the secondary bonds are weak, they are relatively easy to make and break and provide the chains with the means both to stabilize a particular structure and to rearrange it, if required.

Figure 3.3 Parallel chains of nylon showing strong covalent bonds along the chain and weaker hydrogen bonds between chains

Wet, because cells are full of water, and most of the complex processes that operate within them are dependent on that. We tend to take water for granted but it is actually a very remarkable liquid, because the charge associated with the H_2O atoms is not distributed evenly across the molecule. This makes water polar. The oxygen end is slightly negative and the hydrogen ends are slightly positive, so that relatively strong (though still weak!) hydrogen bonds are formed, which hold the molecules loosely together. Because water is polar it doesn't easily mix with uncharged molecules, such as oil, because this would disrupt the hydrogen bonding: on the other hand it has a strong affinity for charged species like ions, or polar side chains on a polymer. Water can therefore have a profound effect on the arrangement of other molecules around it, mixing easily with polar sections but causing non-polar groups to clump together out of the way.

Sticky, because as you'll know from your consideration of scaling in Block 1, the relative effect of different forces at small scales can be quite different from what we are used to in the macro world; electrostatic forces, in particular, are highly significant. While this can cause problems, like the 'stiction' mentioned in Block 1, it can also be used to good effect if the molecular design is right. Short-range electrostatic forces are widely used in nature both to stick things together and to keep things apart.

Another property that becomes much more significant at small scales, particularly in the aqueous environment of a cell, is viscosity. Consequently, the sorts of method we employ for moving things around on the macro scale are no longer appropriate. In fact, diffusion is the main mechanism for movement within cells, relying on the constant buffeting from Brownian motion to push things about. The particles in a liquid (or a gas) are constantly moving around and bumping into their neighbours, and any suspended particles will be pushed around by these collisions. The trouble with Brownian motion is that it is random, so if you need something to move to a particular place by diffusion then the only option is to rely on probability and wait for it to arrive there by chance. This is not quite as bad as it may seem, as we'll explore next; nevertheless, it is no surprise that cells have come up with some other approaches to directed motion, which we'll look at later in the block.

If a small particle is suspended in a liquid or a gas it can be seen to move around randomly as a result of collisions with the atoms or molecules around it. This behaviour is known as Brownian motion.

I've said that atoms and molecules are in constant motion, but how fast are they moving? It's quite straight forward to show that for particles moving around in a gas the average kinetic energy ($\frac{1}{2}mv^2$) depends only on the temperature and is equal to

$\frac{3}{2}kT$, where k is Boltzmann's constant and T is the temperature in kelvin. This expression can give us a good idea of the sort of speeds we might expect.

EXERCISE 3.3

The mass of an oxygen molecule is 32 atomic mass units (amu). Taking $k = 1.38 \times 10^{-23}\ \mathrm{J\ K^{-1}}$ and 1 amu = 1.66×10^{-27} kg, estimate the average speed of an oxygen molecule at room temperature (~293 K).

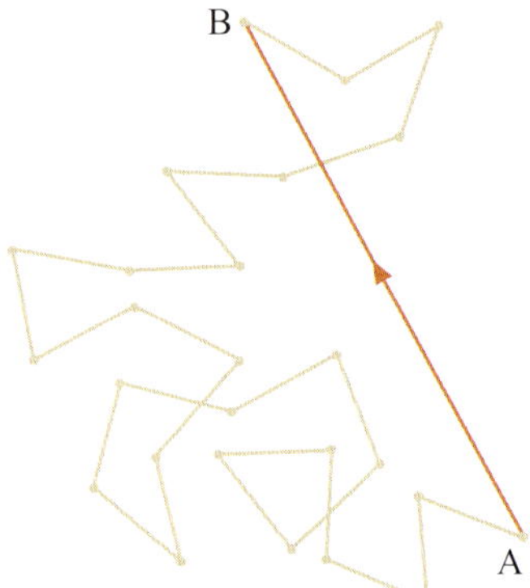

Figure 3.4 A random walk from A to B

Although atoms and molecules may be whizzing around at great speed at the molecular level, they don't get very far in one direction – particularly if they are packed together fairly closely, as in a liquid – because they are constantly bumping into each other. With each collision, the direction of motion changes and the path followed is a succession of zigzag steps called a random walk, as illustrated in Figure 3.4.

There is a statistical rule that tells us that the average distance travelled in a random walk of n equal steps will be $\sqrt{n}$ step lengths. In Figure 3.4, a molecule starting at position A takes 25 random steps of equal length to get to B. Therefore, the distance from A to B is, on average, equivalent to just 5 steps. The distance moved by diffusion, and hence the time taken to get from one place to another, is governed by this mathematical relationship and is crucial to movement on a small scale.

The quantity we use to describe the rate at which one thing can diffuse through another is the diffusion coefficient, D. Roughly speaking, the time, t, it takes a given molecule to diffuse a given distance, x, is given on average by x^2/D. The calculations in SAQ 3.1 will give you an idea of what this means in terms of moving things around within cells.

SAQ 3.1 (Learning outcome 3.1)

(a) Using the diffusion coefficient values below, estimate how long it would take a molecule of:
   (i) oxygen
   (ii) sugar
   (iii) a small protein
   to diffuse across a cell of diameter 20 μm.

   Assume that the cell is full of water.

(b) How long would it take an oxygen molecule to diffuse from one end of your finger to another?

Diffusion coefficient of oxygen in water = $18 \times 10^{-10}\ \mathrm{m^2\ s^{-1}}$

Diffusion coefficient of sugar (sucrose) = $4.5 \times 10^{-10}\ \mathrm{m^2\ s^{-1}}$

Diffusion coefficient of a protein typically $< 1 \times 10^{-10}\ \mathrm{m^2\ s^{-1}}$. You can take this value to represent a small protein.

Because oxygen is moving in a random way, this effectively tells us that an oxygen molecule can cover the entire cell in about a fifth of a second and is likely to get to where it needs to be in considerably less time than that. On this sort of length scale, diffusion powered by Brownian motion is quite sufficient for moving molecules around in reasonable timescales, whether in cells or in any fabricated devices we may wish to design. It is only when the distances increase that other forms of transport need to be considered.

The combination of constant random motion with molecules designed to be sticky in some places and to repel each other in others is a powerful one. Parts of molecules that prefer to congregate together to avoid water have the means to do so, while attractive forces will link other parts together. Any two molecules in the same cell that are designed to fit together will, before very long, find each other and do just that. So it is possible to assemble complex units from their individual components with very little effort, just by getting them together in the same general area. This ability of biological molecules to self-assemble lies at the heart of bionanotechnology.

2 STRUCTURAL MATERIALS

2.1 Biological materials

I'm using the term 'biological material' here to describe materials of natural origin. The more general term 'biomaterial' can include both natural and synthetic materials that are used to replace or interact with part of a living system.

Materials engineers have long recognized the impressive range and combination of properties offered by biological materials. Figure 3.5 shows some representative examples of the combination of tensile strength and toughness (measured by Young's modulus, or elastic modulus for polymers) offered by natural materials, with some more common engineering materials included for comparison.

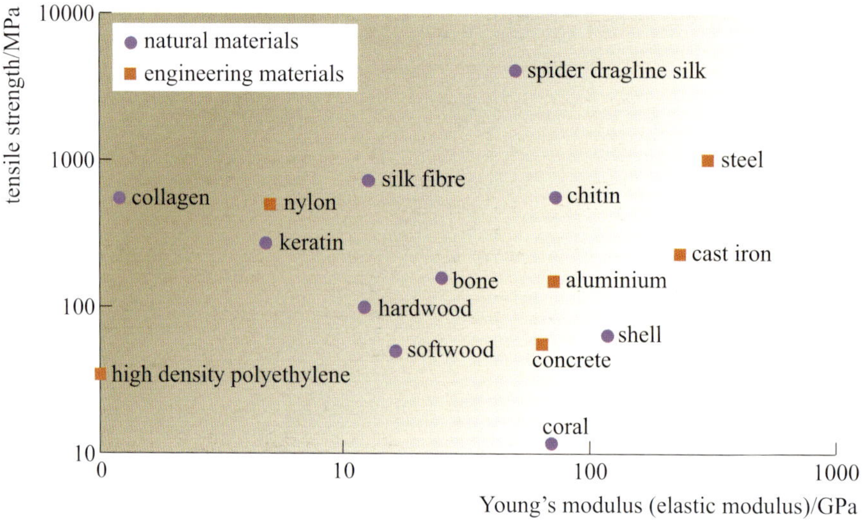

Figure 3.5 Tensile strength and toughness (Young's modulus, or elastic modulus for polymers) of a range of natural materials with some common engineering materials included for comparison. The (diagonal) shading indicates the energy to break: the darker the shading, the higher the value – note the logarithmic axes

Many biological materials approach or exceed the strength of steel: in fact, some forms of spider silk exceed it by a good margin. If density is taken into account, the specific strengths (strength per unit mass) are even more impressive. Biological materials frequently combine high strength with relatively low stiffness, and this leads to high values for the energy needed to break the material, as indicated by the shading on the diagram. You will see that collagen, the main component of skin and tendons, is particularly impressive in this respect. Biomaterials tend to have a specific combination of properties, closely suited to their function, which can be difficult to match with manufactured alternatives.

If we look closely at natural materials we find that their properties are controlled right down to the molecular level, with each material uniquely adapted to suit a particular purpose and to interact in complex ways with its neighbours in response to changes in circumstances. Consider bone, for example. Natural bone is a composite, combining polymeric protein chains with crystals of ceramic calcium phosphate. In the body it responds to stress levels, being absorbed when low stress levels indicate that it is not needed, growing thicker when the stress levels are high and receding in situations where the stress levels are so high that damage could occur. Given this intelligent response, together with the mismatch in mechanical properties of bone and steel evident from Figure 3.5, it is no wonder that the design of effective hip replacements, for example, is so technologically challenging.

One of the reasons that it is so difficult to match the properties of biological materials is that they are built within cells from the bottom up, unit by unit, to a strict specification. If, by studying their structure and method of production, we can learn to copy the design principles used in nature, then the scope for producing new, smarter materials outside the cell is immense.

SAQ 3.2 (Learning outcome 3.3)

A particular spider of mass 50 mg can spin dragline silk with a tensile strength of 1.2 GPa ± 0.2 GPa. This compares to 0.55 GPa ± 0.05 GPa for a fibre of nylon 6 (a common form of nylon). The tensile strength is the force per unit area needed to break the fibre.

(a) What diameter must the dragline exceed to ensure that it can support the weight of the spider?

(b) What diameter fibre would be necessary to support a 60 kg human? What is the comparable value for nylon 6?

(c) Spider silk and nylon 6 have similar densities of about 1200 kg m^{-3}, while that of steel is 7800 kg m^{-3}. Assuming the steel to have the same tensile breaking stress as the spider silk, calculate the mass per metre of a fibre of sufficient diameter to support the 60 kg human, for all three materials.

Several of the materials in Figure 3.5, including collagen, keratin and silk, are proteins. Proteins fulfil many important roles in cells and make up much of their mass. You probably think of proteins in terms of foodstuffs like eggs, fish and meat; in fact, looking at the contents of the food that we eat gives a pretty good guide to the range of materials found in cells. For example, a chicken curry from my fridge lists the following composition (by mass): 15% protein, 9% fat, 5% carbohydrate, 0.8% salt (and the rest mainly water). Compare this to the chemical composition of a typical mammalian cell (shown in Table 3.1) and you'll see that they are remarkably similar.

Table 3.1 Chemical composition of a typical mammalian cell

Compound	*Mass/% of total*
Water	70
Proteins	18
Lipids (fats and oils)	5
Polysaccharides (carbohydrates)	2
Nucleic acids	1
Small molecules and ions	4

Despite their complexity, living cells are constructed almost entirely from four basic classes of chemical compound:

- lipids, which are used for energy storage and to form cell membranes
- proteins, which fulfil a very wide range of functions from structural materials to nanoscale motors
- sugars and polysaccharides, which are important for energy storage and as structural materials in plants

- nucleic acids, DNA and RNA, which although less prolific are hugely important as they allow information to be stored and read within cells.

Lipids, proteins and nucleic acids will all be considered in more detail. In this section, I'll introduce lipids and proteins as the materials that give structure to cells.

2.2 Construction with lipids

The cell membrane shown in Figure 3.1 is constructed from lipids. Chemically, lipids are a rather varied group of compounds that include all the substances you might already think of as fats or oils. What they have in common is that they are all insoluble in polar liquids like water, but soluble in organic (carbon-based) solvents: by this I mean the sort of smelly solvents you tend to find in paints, glues and degreasing agents; chloroform is one example. Lipids make up the fatty components of living organisms and are the major components of margarine, cooking oil, butter and the white fat associated with meat. The terms 'fat' and 'oil' simply describe whether a lipid is solid (a fat) or liquid (an oil) at room temperature. Fats and oils are important in cells as energy storage compounds, while the cell wall itself is constructed from a class of lipids known as phospholipids.

The interplay between hydrophilic and hydrophobic groups is important in many biological materials. Section 11 Weak bonds in the *Principles* wing describes hydrophobic interactions and other important types of weak bond in more detail.

Phospholipids and their structure are described in more detail in Section 12 Phospholipids in the *Principles* Wing.

The reason for their insolubility in water is that lipids contain large hydrophobic groups. *Hydrophobic* literally means 'water-hating' and generally applies to non-polar substances or chemical groups. In contrast, polar substances or groups are termed *hydrophilic* or 'water-loving'. As the name suggests, they have an affinity for water and are more likely to dissolve in it.

Their hydrophobic nature means that lipids tend to separate from water and from all water-soluble, hydrophilic compounds and to group themselves together as large hydrophobic aggregates, held together by weak hydrophobic interactions. This is why oil and water don't mix. Because of the **weak bonding** between the molecules, lipids tend to be fluid aggregates that can easily change shape. Think of a lump of lard: if you press on it, it dents. A material like this with a certain amount of fluidity can be extremely useful in certain situations. Insolubility in water is also a useful property when it comes to forming a barrier to keep some substances separate from others.

Many important lipids, including the **phospholipids** that make up cell membranes, contain both hydrophobic and hydrophilic groups. Such molecules are termed 'amphiphilic' and exhibit particularly interesting properties when mixed with water. Figure 3.6 is a sketch representing an amphiphilic lipid molecule, such as a phospholipid. The two 'tails' are based on chains of carbon atoms and are hydrophobic. The head is a polar hydrophilic group. Other lipids may have a different number of tails.

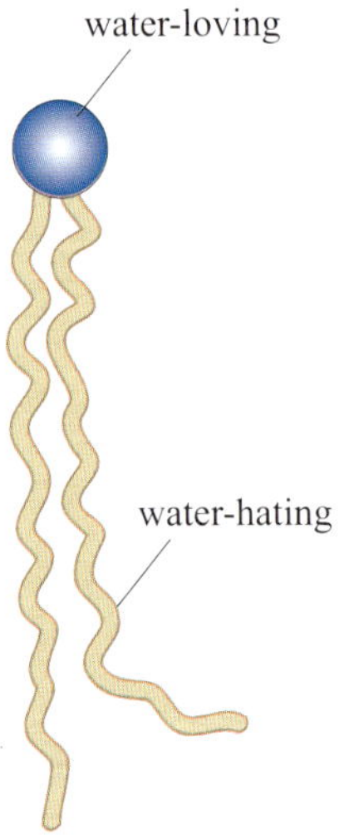

Figure 3.6 Simple representation of a lipid molecule. This one has two hydrophobic (water-hating) tails and a hydrophilic (water-loving) head, like a phospholipid

Consider how amphiphilic lipid molecules might aggregate together in a strongly hydrophilic, watery environment. The hydrophobic chains will avoid water, coming together and interacting with one another on the inside of the aggregate. This leaves the hydrophilic head groups on the outside of the aggregate, in contact with water molecules. The charged head groups will repel each other to some extent, so the arrangement they adopt must achieve a balance between the attractive forces (between the heads and the surrounding water) and the repulsive forces (between neighbouring heads). A wide range of structures can be produced, depending on the shape of the molecule, the nature of the head group, and the concentration of the mixture. Some examples are shown in Figure 3.7.

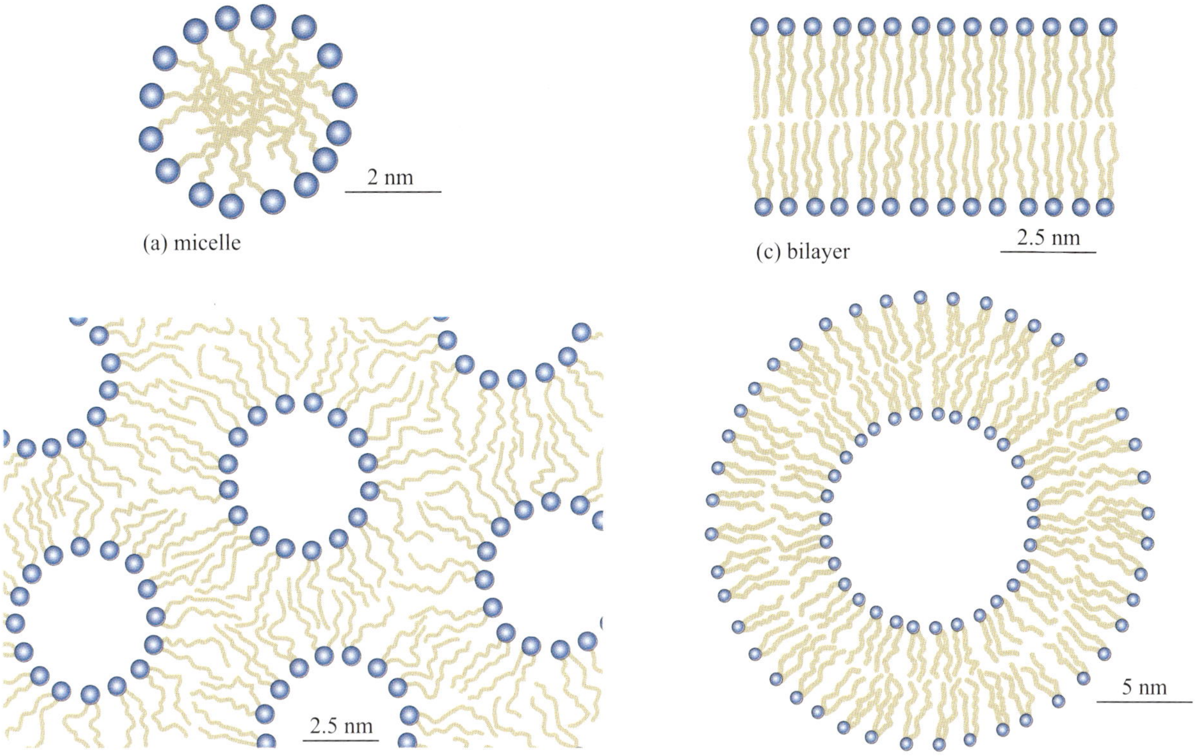

Figure 3.7 Cross sections of some of the common structures adopted by amphiphilic lipids in water: (a) micelle; (b) inverted micelles; (c) bilayer; (d) bilayer vesicle, or liposome. The formation of these structures is explored further on the course DVD

The formation of well-defined structures by amphiphilic lipids is the first of several examples we will see of *self-assembly*, which is a key feature of many natural systems. Self-assembly means that no outside influence is needed to cause the molecules to organize in a particular way: it is a direct consequence of their molecular design and the electrostatic forces that exist, on very small scales, between individual molecules. It relies on Brownian motion to shake the molecules around so that the optimum arrangement can be found.

Amphiphilic lipids, which contain a polar, water-soluble group attached to a non-polar, water-insoluble hydrocarbon chain, have many uses. They are widely used as surfactants (substances that reduce the surface tension of a liquid) and emulsifiers (substances that promote the formation of a stable mixture of oil and water), and you'll find plenty of examples around the home. Soaps and detergents work by forming micelles when added to water; see Figure 3.7(a). These micelles are able to surround non-polar contaminants such as grease, encapsulating them within their hydrophobic core and thus allowing them to dissolve in water, as illustrated in Figure 3.8. The charged heads on the surface ensure that the micelles repel one another, preventing them from clumping together – effectively emulsifying the dirt. In other circumstances (higher concentrations, or differently shaped lipids) the micelles may be inverted, as in Figure 3.7(b). In this case the hydrophilic heads point inwards, surrounding small globules of water within a matrix of lipids.

Lipids are also common in the kitchen. Many foodstuffs contain both oil and water, and a variety of emulsifiers are used to stabilize their structures. For example, lipids found in egg yolk act as an emulsifier in egg mayonnaise, enabling a stable emulsion to be formed from oil and vinegar.

Some authors would describe the behaviour of lipids as *self-organization*, since there is more than one possible way of arranging the molecules, and reserve the term *self-assembly* for more specific interactions such as those found in proteins. Here, I will not differentiate between the two and only the term self-assembly will be used.

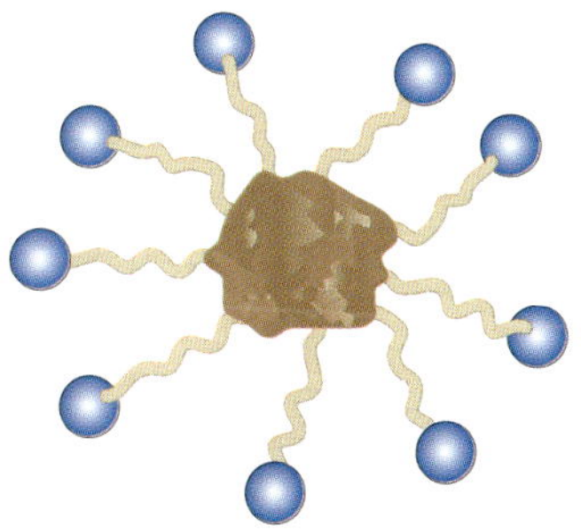

Figure 3.8 Dirt surrounded and made soluble by lipids

Plant cells possess a tough outer cell wall in addition to the phospholipid membrane. Here we will concentrate only on the membrane itself.

The double-layered lipid structure shown in Figure 3.7(c) illustrates the basic arrangement of the phospholipids in a cell membrane. A more detailed view is given in Figure 3.9. Within the membrane the phospholipids cluster together to form a sheet-like structure, with all the hydrophobic tails in the centre, screened from the surrounding water. The phospholipid bilayer of a cell membrane is a very effective barrier, preventing hydrophilic molecules from passing freely into, or out of, the cell. Membranes also define the boundaries of structures within cells, restricting various cellular functions to particular compartments of the cell. Another important function of the membrane is to support other molecules, which allow the cell to interact with its neighbours and with the external environment. Larger molecules, such as proteins, 'float' within the fluid lipid bilayer; we'll return to consider these later.

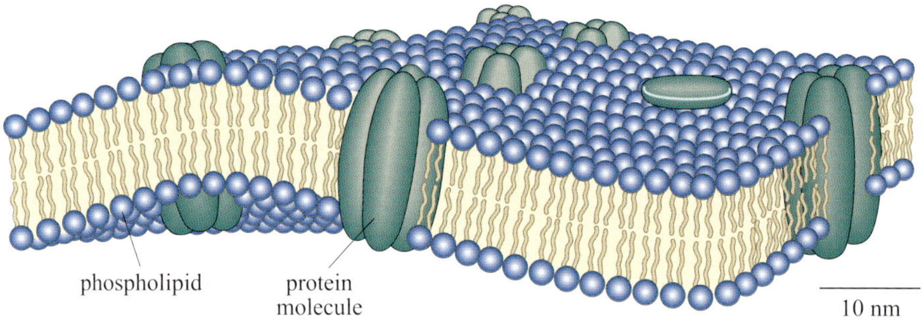

Figure 3.9 A biological membrane; a typical membrane thickness is about 10 nm. The green structures are membrane proteins; some of these traverse the membrane and others are on its surface

The term liposome tends to be used for synthetically produced bilayer structures. The term vesicle is used more generally, but has a specific meaning within cells as a small spherical bilayer sac in which substances are transported or stored.

The final structure illustrated in Figure 3.7 shows a lipid bilayer bent round on itself to form an enclosed compartment. These structures may be described as liposomes or vesicles, depending on the context. The bubbles formed by soaps and detergents also have a bilayer structure similar to that in Figure 3.7(d) and provide us with a reasonable model for a cell membrane. Blowing a bubble shows us how a flexible membrane can be bent round on itself to form an enclosed space; you can demonstrate the fluid and self-sealing nature of a membrane by moving an object, such as a pencil, through a soap film and then removing it.

Liposomes can be used to encapsulate water-soluble substances such as vaccines, drugs, enzymes or vitamins and deliver them to different cells in the body. Drug-delivery systems based on lipids are under development and show considerable potential. Elsewhere, aggregates of lipids in solution are being used to produce nanoparticles of precisely controlled size and shape, such as the quantum dots available for molecular labelling. ▶***Langmuir–Blodgett films***◀ (named after the scientists who first studied them) provide a well-established method of generating structures built up from lipid monolayers that have potential for electronic applications.

Synthetic bilayer sacs like those in Figure 3.7(d) are relatively easy to make and provide a reasonable model for a (very!) simple cell. But for many nanoengineering applications it is much more useful to have a lipid layer that is tethered in a fixed position. Fixed bilayer membranes supported on solid surfaces are much more stable and often offer a more promising route for the development of structures that mimic nature.

In related developments, the self-assembling properties of lipid-like molecules containing thiol groups have been used in microcontact printing and dip-pen lithography to produce self-assembled features on gold surfaces at scales as small as tens of nanometres. By varying the chemical structure of the molecules used, the surface chemistry can be modified so that the thiol may function as an etch resist or produce a surface with markedly different properties in different areas.

Thiol groups, which contain sulphur and hydrogen, are particularly useful for 'sticking' biological molecules to metal surfaces. They are introduced in Section 11 Weak bonds in the *Principles* wing.

▶Langmuir–Blodgett films◀

A Langmuir–Blodgett (LB) film is made up of highly organized layers of lipids, each just one molecule thick, deposited onto a solid substrate by repeatedly dipping it into a water bath with a lipid film floating on top. By varying the experimental conditions and the lipid used, multilayer stacks can be built up with precisely controlled molecular compositions. LB films may consist of a single layer or many, up to a depth of several micrometres, and can be made with various electrochemical and photochemical properties. They are being investigated as possible structures for tiny integrated circuits.

SAQ 3.3 (Learning outcome 3.2)

(a) What do the four structures in Figure 3.7 have in common?

(b) If salt (sodium chloride, Na^+Cl^-) were added to the mixture in Figure 3.7(a), where would it go? What about petrol (which contains hydrocarbon chains, such as C_8H_{18})?

SAQ 3.4 (Learning outcome 3.2)

Suggest which of the structures shown in Figure 3.7 would be suitable for the following applications:

(a) restricting the size of crystals growing from solution in water

(b) suspending a hydrophobic pigment molecule in a water-based paint

(c) providing a boundary between an area of high concentration and an area of low concentration within a liquid

(d) delivering a water-soluble vitamin to a cell.

The principles that govern the self-assembly of lipids can also be applied if we take small units with different chemical characteristics and join them together. This is the situation that exists in a block copolymer, as described in ▶***Self-assembly in block copolymers***◀.

►Self-assembly in block copolymers◄

Block copolymers have chains made up of alternating sequences of two or more different repeating units. Let's assume that there are just two components, which I'll call A and B, so that a section of chain might look something like this:

AAAAAAAAAABBBBBBBBBBBBBAAAAAAABBBBBBBBBBBBBB

Now suppose that the A units are hydrophilic and the B units are hydrophobic. The As will try to separate from the Bs and clump together, and vice versa, leading to phase separation on a tiny scale. The structures that can be produced are very similar to those obtained from lipids but, in this case, the result is a solid, with greater connectivity throughout the structure because the units are linked together in chains. Figure 3.10 shows the sort of structural arrangements that can be obtained. By changing the length and proportions of the different blocks the microstructure, and hence the bulk properties of the material, can be finely controlled.

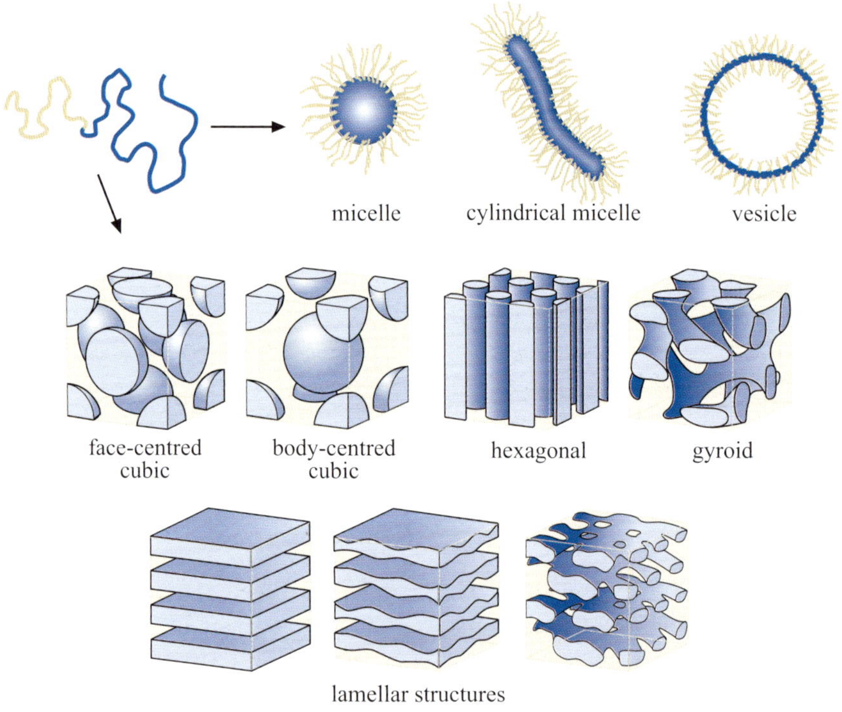

Figure 3.10 Examples of the different microstructures adopted by block copolymers with hydrophobic and hydrophilic elements

2.3 Providing a framework: structural proteins

Of course, our bodies can't just be made up of squidgy bubbles of phospholipid, or we would collapse in a heap on the floor! Stiffer frameworks, both inside and outside the cells, also exist and help to define shape and add strength. These frameworks are formed largely from *structural proteins*, a class of polymeric materials that form fibres and filaments to provide mechanical support for cells and tissues. Structural proteins are made inside cells but are often then moved into the space surrounding the cells, where they interact together to form a three-dimensional polymer network, permeated by fluid.

Section 2.2 introduced the idea of self-assembly in a simple system, but the real masterpieces of self-assembly are the **proteins**. As well as providing structure, proteins play a central role in defining the function of different cells. There are many of them – around 10 000 types in a typical cell – and they fulfil a wide range of functions. Several of these functions will be covered in the block and an overview is provided in the section on protein diversity in the *Principles* wing.

Section 13 Proteins in the *Principles* wing gives an overview of the wide range of roles carried out by proteins in cells.

Proteins are one example of a *biopolymer*. You will already be familiar with synthetic polymers such as polyethylene and nylon: long chains made up of many thousands of repeating units, called monomers, linked together by strong covalent bonds. Polymers are particularly versatile materials because of the very different strengths of the bonds between monomer units in the chain (strong) and between one chain and another (weak). By varying the arrangement of chains within a material a huge range of properties can be achieved, from extended chain fibres as strong as steel, to soft rubbery networks where the chains are randomly coiled together, or rigid networks where cross-linking keeps every chain tightly pinned in place. The same principles apply to biopolymers, although the chemical composition of the chains tends to be rather more complex.

Much of the functionality of a protein depends on the precise three-dimensional shape taken up by the molecule, an idea that will be explored further in the *Principles* wing. Structural proteins such as collagen tend to be long and thin, like the strands in a rope: such proteins are collectively known as fibrous proteins.

Despite the huge number and diversity of proteins in cells, they are all based on the same relatively small number of fundamental building blocks: 20 different amino acids. Amino acids are a class of compound with the same basic chemical structure: their chemistry is described in more detail in the *Principles* wing, while Figure 3.11 offers a more diagrammatic representation.

Figure 3.11 Diagrammatic representation of an amino acid

At the centre of each amino acid is a carbon atom, which forms bonds to four other groups. Three of these groups are always the same. Two of them can react together to link individual units into a chain, while the third common group is a hydrogen atom. What distinguishes one amino acid from another is the fourth group, known as the R group, shown as a hook in Figure 3.11. There are many possibilities for R, but only 20 of them are found in proteins in nature, at least on this planet. Figure 3.12 shows a few amino acid units linked together, each with a different R unit.

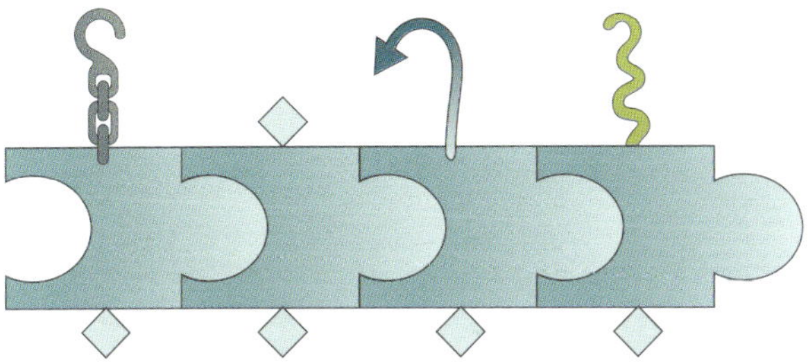

Figure 3.12 Diagrammatic representation of four different amino acids linked together

The amino acid units are linked together along the chain by strong covalent bonds, which I have represented in Figure 3.12 by tightly interlocking pieces. The bonds formed are exactly the same as those that link the units together in synthetic nylons, although (for no particularly good reason) the terminology used is different. In nylons, the monomers are usually referred to as amides, the polymer as a polyamide, and the bonds linking the monomer units together as amide bonds. Biochemists are more likely to refer to amino acids, polypeptides (the protein chains) and peptide bonds.

Why have I represented some of the R units by hooks? As shown in Figure 3.12, the R units protrude from the side of the polypeptide chain. Some of them are very small (e.g. a hydrogen atom) while some are much larger. The side groups play an important role in the behaviour of a protein because they can form weak bonds with other units, and help to stabilize particular folded-chain arrangements. The different hooks are intended to show that the side units can link together in different ways, but that these links are much weaker than the links along the chain.

Protein chains may be many hundreds of units long, so the number of possible sequences is enormous. The thousands of different proteins found in cells represent just a small fraction of these possibilities; they have evolved over time to perform specific biological functions. Some amino acids may not be used at all in a particular protein, whereas others may occur many times. Each different protein is unique because it has its own individual sequence of amino acids along its length. The sequence of amino acids that makes up a particular protein chain is known as the *primary structure* of the protein.

Biochemists use the terms ‘protein’ and ‘polypeptide’ rather loosely and interchangeably, and both terms will be used here as well. When a protein is synthesized in a cell, the linear polypeptide chain folds as it is produced. The biological activity of this newly formed molecule depends on its three-dimensional, folded structure, and when considering a polypeptide chain as a three-dimensional structure, it is generally referred to as a protein.

The sequence of amino acids along a polypeptide chain determines the primary structure of the protein, but the chain is flexible and has the potential to fold up in different ways. In practice, only a very small proportion of the possible folding patterns are energetically favourable, stabilized by interactions between the ‘hooks’ protruding from the main chain. The spatial arrangement, or *conformation*, adopted by a chain is known as its *higher-order structure* and different levels of organization can be identified. Indeed, one of the remarkable things about proteins is that any chain with the same sequence of amino acids will fold up in exactly the same way, provided the conditions are the same, although a bit of help is sometimes needed to achieve this. Thus primary structure determines higher-order structure, which in turn, as we shall see later, determines function.

Collagen is a particularly abundant fibrous protein that accounts for 25% of all body protein. It has high tensile strength and is found particularly in skin and tendons. The higher-order structure of collagen is illustrated in Figure 3.13. The basic structural unit is a triple helix, composed of three polymer chains (which are themselves helical) coiled round each other and held together by weak, non-covalent bonds, as shown in Figure 3.13(a). These units are then packed side by side in small groups of up to 200 to form larger fibrils, Figure 3.13(b), which are bundled together to form fibres, Figure 3.13(c). The fibres may themselves clump together in larger aggregates to build up body tissues. The result of all this organization is a very strong, reasonably stiff, tough material.

One of the most powerful methods available for working out the structures of proteins is X-ray diffraction: a technique that relies on the scattering of X-ray radiation by a regular crystal lattice. It is not always easy to persuade a protein to crystallize, but as

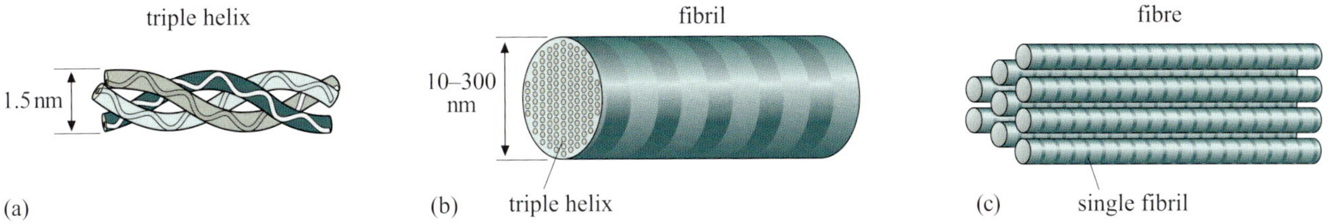

Figure 3.13 A triple helix forms the basic structural unit of collagen (a), up to 200 of which are packed together to form fibrils (b), which themselves pack together to form fibres (c)

techniques improve and higher-intensity X-ray sources and more powerful analytical tools become available, more and more proteins are being characterized in this way. Computer modelling is then applied to translate a diffraction pattern to a likely molecular structure. The *Protein Data Bank (PDB)* is a website maintained by the Research Collaboratory for Structural Bioinformatics (RCSB), which acts as a central repository for results in this field and provides an excellent example of international cooperation between scientists. Structures for most of the proteins mentioned in this block, and interesting articles about many of them, can be found on the PDB website.

Collagen provides us with a good example of how protein molecules can self-assemble to produce a structure well suited for a particular purpose. It can also be used to demonstrate the limitations of self-assembly. The three intertwined strands that make up the basic unit of collagen are created and assembled simultaneously within the cell. If we separate the strands – which we can do very easily by heating collagen in water above 70 °C – and then allow them to re-assemble, the triple helices will not re-form. Instead, separate strands interact with one another to form many links with other molecules and the result is a large, open network. This is exactly what happens when you make a jelly. Breaking down the natural structure of a protein is known as *denaturing*. The individual molecules from the denatured collagen are what we call gelatin. When a jelly sets, the chains link together to form a huge, continuous, three-dimensional network like a giant cage, which traps the water inside through hydrogen bonding to the protein molecules.

EXERCISE 3.4

Consider a short segment of protein chain, just six monomers long.

(a) What is meant by the primary structure of the chain?

(b) How many different types of monomer are available in nature?

(c) How many different sequences are possible in the six-monomer segment? Assume that all the possible monomers are available at each of the six positions.

Regular helical structures like those that occur in collagen are very common in proteins, because they allow many weak bonds to form and these stabilize the structure. Figure 3.14 shows a schematic representation of two particularly common regular folding patterns, the α-helix and the β-sheet. You can examine how these structures are stabilized by hydrogen bonding, and see how they can be used to create larger features, in Section 13 Proteins in the *Principles* wing.

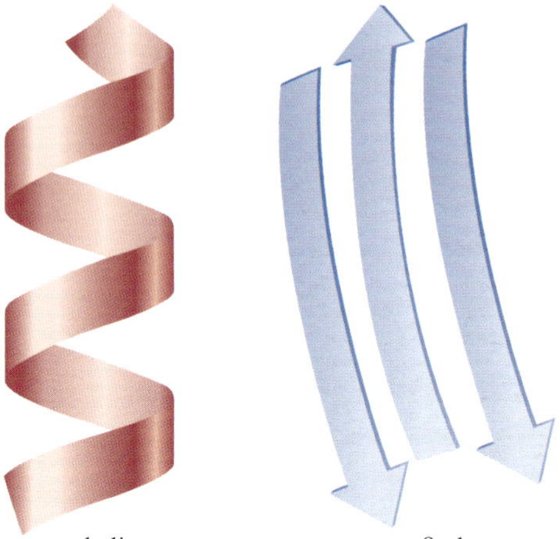

Figure 3.14 Schematic representation of α-helix and β-sheet structures

Keratin is another important example of a fibrous protein, with a different amino acid sequence. It can exist in different forms, based either on α-helices (in mammals) or β-sheets (in birds and reptiles). It is a major component of skin, hair, nails, hooves, horns, claws, beaks, feathers and scales, which gives a good indication both of its versatility and its strength.

More disordered network structures can also be produced, by incorporating segments in the polypeptide chain that can form cross-links to neighbouring chains. This results in a rubber-like structure with high extensibility; elastin, a key component of skin, has this structure.

The presence of regions of helical and sheet-like structures within a protein will affect its properties in different ways: a particularly striking example of this is provided by ►*spider silk*◄.

►Spider silk◄

A polymer that forms fibres stronger, by mass, than steel and can be processed from water at ambient temperature and low pressure. As a consequence of its biological origin, it is extremely environmentally friendly and totally recyclable. It may sound like science fiction, but this is exactly what we have in spider silk. Mind you, it has taken 400 million years to develop!

Spider silk is composed of proteins. Spiders make their webs and perform other tasks using up to seven different types of silk fibre. Each silk has distinct mechanical properties, which arise directly from the primary and higher-order structure of the proteins from which they are constructed. Table 3.2 lists three types of silk from the golden orb web-weaving spider (*Nephila clavipes*) shown in Figure 3.15 and compares their mechanical properties to some synthetic polymer materials.

The dragline silk, which forms the main structure of the web and also supports the spider when climbing or dangling on a thread, is exceptionally strong and reasonably elastic, making it a very tough fibre comparable to the best that modern polymer technology can offer. By contrast the flagelliform silk of the spiral part of the web is not only sticky but very stretchy, so that it can absorb the energy from a collision with a large insect without breaking, and keep the insect trapped in the web. The silk from the minor ampullate gland is as strong as the flagelliform silk but has much lower elasticity and is used for structural reinforcement.

Figure 3.15 Golden orb spider on its web

Table 3.2 Comparison of mechanical properties of spider silk and other polymer fibres

Material	*Use*	*Tensile strength /MPa*	*Extension at breaking point/%*	*Energy to break /kJ kg^{-1}*
Dragline silk	web frame and radii, support when climbing or falling	4000	35	100
Flagelliform silk	core fibres of adhesive spiral	1000	>200	100
Silk from the minor ampullate gland	web reinforcement	1000	5	30
Kevlar	high-strength applications	4000	5	30
Rubber	high-elasticity applications	1	600	80
Nylon 6	wide range of uses	70	200	60

(Source: adapted from Hinman H. B. et al. (2000) 'Synthetic spider silk: a modular fiber', *Tibtech*, vol. 18, no. 9, pp. 374–379.)

Not surprisingly, given its impressive properties, spider silk has been the focus of much research. Scientists have identified the major proteins that make up the different silks and have mapped their amino acid sequences. It turns out that there are just a few key repeating patterns, which in turn lead to particular secondary structures. These are summarized in Table 3.3: the letters G, P and A represent the amino acids glycine, proline and alanine respectively, with X as a 'wild card'. There are two helical structures present here, known as a β-turn spiral and a 3_{10} helix; you can think of these as acting like a loose and a tight spring. The structures also include spacers to separate the repeating units into clusters, but these aren't included in the chart.

Table 3.3 Structural modules found in spider silk from the golden orb spider; read across the rows to see the composition of each type

		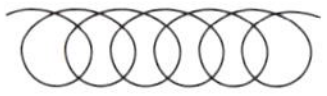 *Elastic β-turn spiral*	 *Crystalline β-sheet*	 3_{10} *helix*
Amino acid sequence		GPGXX	$(GA)_n$ or A_n	GGX
Dragline silk	protein 1		✓	✓
	protein 2	✓ (~5 repeats)	✓	
Flagelliform silk		✓ (~50 repeats)		✓
Silk from the minor ampullate gland	protein 1		✓	✓
	protein 2		✓	✓

(Source: adapted from Hinman H. B. et al. (2000) 'Synthetic spider silk: a modular fiber', *Tibtech*, vol. 18, no. 9, pp. 374–379.)

The correlation between the secondary structures within the proteins and the physical properties of each silk is striking, as you'll see by working through SAQ 3.5.

SAQ 3.5 (Learning outcome 3.3)

Use Tables 3.2 and 3.3 to answer these questions.

(a) Which two types of silk have the highest extensibility?

(b) What structural feature do they have in common?

(c) To what would you attribute the difference in extensibility between the two?

(d) Why would you expect the β-sheet arrangement to convey strength?

(e) Suggest why a particularly regular amino acid sequence might be needed to form a β-sheet.

(f) How might you account for the high strength but low extensibility of the silk from the minor ampullate gland, by considering its structure?

It isn't just the composition of spider silk that is of interest to technologists, but also the spinning process itself. The formation of secondary structural motifs such as those described result in a liquid crystalline solution, which reduces the amount of force needed by the spider to achieve high orientation. The fibres are extruded through a spinneret in the spider's abdomen (see the cover of this book), which involves a novel internal stretching process not seen in current industrial fibre processing.

The very close link between molecular-scale structure and large-scale properties seen in spider silk suggests that by carefully controlling the sequence of monomer units along a polymer chain we should be able to design materials with very closely specified properties. Indeed, polymer chemists have been doing this in a crude way for many years by designing block copolymers for specific applications. Synthetic block copolymers typically contain only two or three different monomer units, but a wide range of properties can be achieved. The potential degree of control over properties in a polypeptide chain with 20 different monomer options is much greater.

2.4 Engineering with proteins

What are the prospects for designing and making new proteins for specific purposes? The technology exists to build polypeptide chains unit by unit in a test tube, but this is time-consuming and expensive. Often a more practical approach is to find ways of working with nature to produce useful substances in a form that we can use. This might involve extracting a naturally occurring protein and chemically modifying it in some way, or using genetic engineering to produce a particular protein in controlled conditions. An example of how this has been applied to spider silk is described in ▶ *Genetic engineering: spider silk from goats?* ◀.

▶ Genetic engineering: spider silk from goats? ◀

Proteins are made in cells by following coded instructions carried by the genetic material, DNA. A gene is a portion of DNA that holds the instructions for a particular protein. If these instructions are changed, or added to, then different proteins can be made. Genetic engineering is the term given to a range of new research techniques that allow us to manipulate the DNA in cells.

Gene-splicing is a method for transferring selected genes from one species to another, by cutting a fragment from the DNA of one organism and joining it to a DNA molecule from another organism. Bacteria are often used as the host and can be used to produce large quantities of artificial proteins in the lab.

Spider silk can't easily be made in bulk by farming spiders, because they have an unfortunate habit of eating each other. Transferring the gene to bacteria has not proven successful either, producing silks that are impossible to spin or have low strength. Consequently, some researchers have turned to mammal cells as a possible alternative host and, in 2002, a Canadian company announced that it had succeeded in breeding goats carrying a modified version of the spider-silk gene. The protein can be extracted from the goat's milk, and spun into yarn in a process that mimics conditions in the body of the spider.

The technology is still a long way from practical production, but if development continues we could see spider silk rivalling synthetic polymers like nylon and Kevlar in specialist applications where high toughness is required, from bullet-proof clothing to reinforced composites for aircraft panels.

Using naturally occurring proteins in new ways is one thing, but the ultimate goal of designing new protein-based materials or devices from scratch is much, much more difficult. I chose spider silk as an illustration of structure–property relationships within proteins because it has been widely studied and the links between the conformation of the protein chain and the mechanical properties of the material are, at least at a superficial level, relatively easy to spot. However, for most proteins, particularly those with more complex functions, such links are much harder to establish.

Designing a protein does not just involve choosing the right ingredients to achieve a desired structural element: it must also take account of possible unforeseen interactions which might lead to unwanted folded-chain conformations. This aspect is often termed *negative design*. The goal of negative design is to ensure that only one folded-chain conformation is formed, by ensuring that all other possible structures are energetically unfavourable. Not a trivial task!

In one simple example of protein design, scientists are working to produce improved materials for making water-based gels (hydrogels), in a similar way to that in which denatured collagen associates to form jelly. Hydrogels have many important uses, from highly absorbent dressings and water-retaining granules for agriculture, to structural devices such as contact lenses and components for microfluidic analysis systems.

The polypeptide chains designed to make the gels in my example consist of two different modules. The central portion is a long, disordered chain that cannot fold into a regular structure, but that includes hydrophilic side groups that can associate with water. At the ends are stretches of chain with a row of hydrophobic side groups. The groups from one chain can line up side by side with those from another in a zipper-like arrangement known as a leucine zipper. These form the cross-links, which are weak enough to break apart when heated. By varying the length and the precise composition of the different sections, a range of proteins with different gelation properties can be made. Leucine zippers are described in more detail in Section 13 Proteins in the *Principles* wing.

However, in most cases we are a very long way indeed from being able to design a protein from scratch for a particular function. At the moment, the best we can typically hope to do is to identify the function of naturally occurring protein-based molecular devices and try to identify the key parts of the molecule and how they work. Then, if we are very lucky, we might get away with a tiny bit of tweaking in order to make the molecule fit in better with our system. Many of the current practical applications of bionanotechnology are aimed at doing just that.

SAQ 3.6 (Learning outcome 3.4)

(a) What are the principal design requirements for a protein chain if it is to form a random, cross-linked network capable of trapping water to form a gel?

(b) Which aspects would you classify as positive design and which as negative design?

(c) What extra feature is required if the gel is required to break up on heating?

As an example of how a naturally occurring protein can be put to a new use, and modified for the purpose, let's look at green fluorescent protein (GFP), which is found in certain types of jellyfish. As its name suggests, GFP emits green light, using energy absorbed from ultraviolet light. Controversially, it has been introduced into animal cells through genetic engineering, to create fluorescent varieties of plants and animals, such as the 'GFP bunny' described as a 'transgenic artwork' by its creator, the artist Eduardo Kac. Such uses have sparked important public debate on both the safety and the morality of genetic engineering: issues that are likely to increase in significance as our knowledge and capabilities advance further.

A much less controversial application of GFPs is their use in scientific and medical research to 'label' an object of interest and to study its behaviour. For instance, GFP can be attached to another protein, whose movement around a cell can then be tracked simply by watching the green glow move around under a microscope. Another recent development is to modify GFPs to act as biosensors, which fluoresce in a characteristic way if a particular substance is present.

The structure of GFP is illustrated in Figure 3.16. This type of picture is known as a 'ribbon representation' as it uses ribbons to pick out key structural features. You should be able to spot a β-sheet, here folded up to produce a 'can'. There are also short sections of α-helix, coloured red in the diagram, one of which passes down the middle of the can. In the middle of the β-can, safely protected from polar water molecules that would interfere with its action, is the light-emitting centre, shown in yellow on the diagram. This is made up from a particular sequence of three amino acids, which interact together to form a structure in which the band gaps are suitable for the absorption and emission of light.

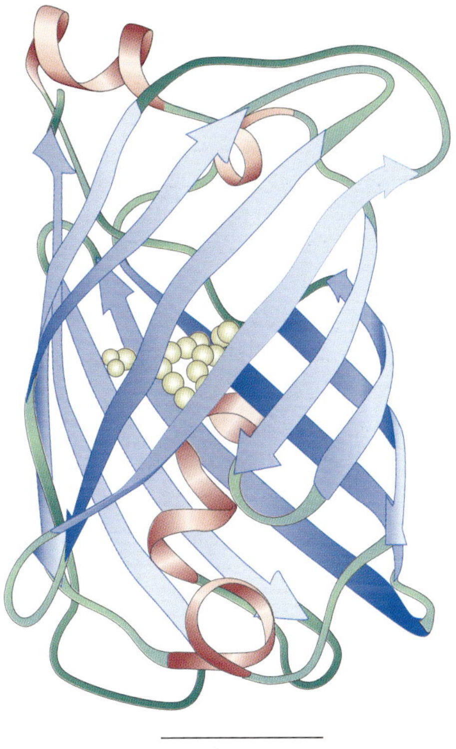

Figure 3.16 Structure of green fluorescent protein: ribbon representation

We can modify this structure by making slight changes in the amino acid sequence or by adding portions to the end of the chain, while keeping enough of the original arrangement to ensure that the basic design is not compromised. How might we want to change it, to increase its usefulness? One possibility is to tweak the colour of the light emitted, by making slight adjustments to the environment of the light-emitting structure. Variations of GFP that fluoresce blue, cyan and yellow have all been made, allowing different substances to be tagged and monitored at the same time. Another common modification is to provide attachment sites for different molecules by adding flexible linkages to the ends of the polypeptide chain, so that the range of substances that can be labelled is increased. In another example, a blue fluorescent protein has been developed to detect the presence of zinc ions; zinc ions interact with the light-emitting centre in a way that increases the level of fluorescence, so that the level of zinc present can be measured from the brightness of the signal.

This is just one example of how we can learn from nature by taking a naturally occurring molecular machine and extending its function to suit our own goals. As we work our way through the cell we will see many more biological devices and learn further principles for working at the nanoscale, which we may be able to adapt and apply to new situations. One day we will surely know enough to design novel nanomachines of our own.

SAQ 3.7 (Learning outcomes 3.3 and 3.4)

(a) Describe aspects of the primary and higher-order structure of the GFP protein that make it particularly suited for its purpose as a light emitter.

(b) Which parts of the structure would you expect to be most open to modification, without compromising the function of the molecule?

EXERCISE 3.5

Suggest an engineering application that could benefit from the manufacture of protein-based structural materials.

3 COMMUNICATION

3.1 Sensing

This section considers how cells detect signals, and how we can apply similar principles to the development of highly sensitive sensors for operation outside the cell. Cells aren't isolated entities but are sensitive to changes in the environment around them; for this they need sensors suitable for all the different stimuli they are likely to receive. Biological sensors can be remarkably sensitive. For instance, specialized cells in the retina of the human eye are capable of detecting individual photons of light if the conditions are right, and our sense of smell allows us to detect many substances at dilutions of less than one part in several billion parts of air. A carrot fly, so my gardening books tell me, can smell a carrot from half a mile away!

The communication methods used in the macro world tend to rely largely on the transmission and detection of vibrations that have sufficient intensity to be distinguishable from the background noise. As we approach the nanoscale, where constant random motion is the norm, such methods become increasingly problematic. You have already seen some examples of small-scale sensors in the previous blocks, so you will know that we can measure physical quantities such as force and rate of rotation with a high degree of precision. But how can we detect a carrot at a distance? And more importantly, how can we distinguish the carrot's signal from all the other stimuli that are constantly bombarding us? For this we need something different; nature has developed sophisticated chemical signalling systems based on molecular messengers, often accompanied by electrical activity. While carrot detection may not be the top priority in engineering development, by understanding how communication works in cells we open up the possibility for developing extremely sensitive and highly specific sensors, for example to identify toxic or explosive chemicals, pollutants, harmful bacteria and viruses, or early warning signs of disease.

If we categorize any device that combines some sort of physical or chemical transducer with a biologically sensitive component as a 'biosensor', then many practical examples are available. Figure 3.17 shows two possible designs, which I will use to illustrate the general principles. Figure 3.17(a) is a simple design based on a cantilever, like those you met in Block 1, while Figure 3.17(b) is a more complex design that has more in common with the sensors found in living cells.

Both the biosensors illustrated in Figure 3.17 rely on the use of a 'capture probe' capable of interacting with the substance to be detected, which I've called the 'target molecules'. The arrival of target molecules can be detected by a change in the

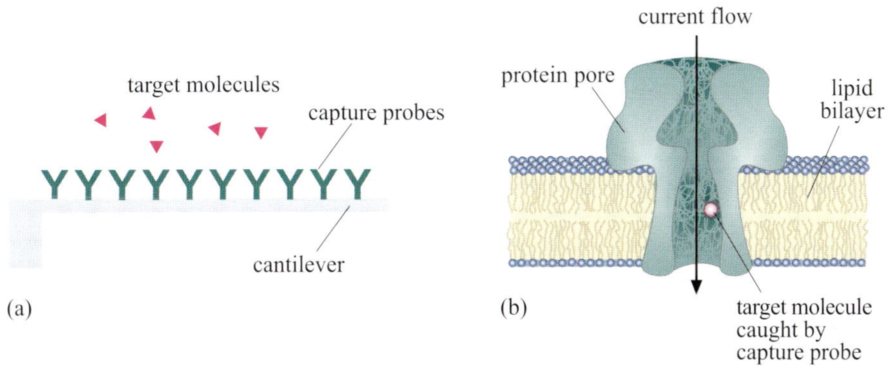

Figure 3.17 Examples of biosensors based on (a) a cantilever and (b) a protein pore

mechanical behaviour of the cantilever in Figure 3.17(a), or by modulation of the current flowing through the pore in Figure 3.17(b). But first let's consider how the target is captured. The capture probe used in a biosensor is often a protein molecule or a fragment of DNA, since both are highly selective and interact only with very specific reagents. For now I'll concentrate only on proteins: you'll find out more about DNA in a later section. To understand how a protein can act as a highly selective detector, we need to look more closely at the relationship between protein structure and function.

3.1.1 Sensing with proteins: binding sites and ligands

Section 13 Proteins in the *Principles* wing provides examples of the range of functions that can be performed by proteins, and also a few of the fantastic shapes adopted by protein molecules and the molecular devices built from them. Much of the functionality of a protein depends on the precise three-dimensional shape taken up by the molecule. Structural proteins, as you have seen, tend to be long and thin, like the strands in a rope. The molecules they are based on may be stretched out to form fibres, like collagen, or they may be built up from individual units in which the chain is coiled up rather like a ball of string: actin, which you'll meet later, is an example of this. However, proteins with more specialized roles, such as the receptor proteins involved in sensing, tend to have a more globular shape.

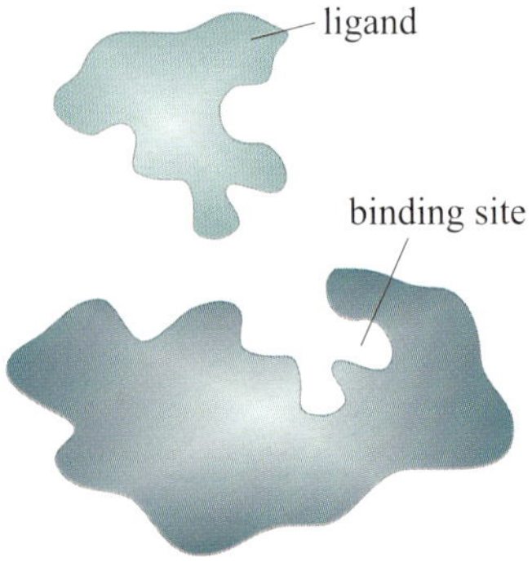

Figure 3.18 A ligand fits into a binding site on the surface of a protein

As with fibrous proteins, the primary and higher-order structure of a globular protein will be critical in determining its function, and α-helices, β-sheets and structural elements composed from them are common. Here we encounter a new principle, too. Proteins have binding sites on their surfaces. These are indentations with a characteristic shape and charge distribution such that only a very specific component can fit into them and, on fitting, induce a change. To use a protein as a detector we need to find (or try to make?) a protein with a binding site which has exactly the right shape and charge distribution for the substance we want to detect to fit into, and attach that protein to some sort of transducer that allows us to detect a change when this happens. In biology, the molecule that fits into a binding site is called a ligand, as illustrated in Figure 3.18.

Within a cell, transduction of the signal carried by the ligand is achieved through shape changes. The act of attaching a ligand at one binding site on a protein may cause the shape to change slightly elsewhere. As a consequence, a channel may open through the protein, other binding sites may appear on the surface of the protein so that additional ligands can attach, or existing binding sites may become less effective and release their ligands. So a small change at one point on the surface of the protein can induce a different change elsewhere. Globular proteins fulfil a wide range of very specific functions, but all rely on this basic principle of binding sites and ligands: you will meet several examples as you work through this block.

In synthetic biosensors a wider variety of methods are employed to detect and relay a signal.

3.1.2 Cantilever-based sensors

Having identified a suitable protein to act as a capture probe, a mechanical sensor can be constructed by applying this protein to the surface of a cantilever of appropriate spring constant. Silicon is a popular choice for the cantilever material, often coated with gold to provide a good surface to attach the capture probes to. The cantilever may be used in stress-sensing or mass-sensing modes. There are, therefore, two sensing processes available in this sensor; this is very common in biosensors. For stress sensing, only one side of the cantilever is coated with capture probes; let's assume this is the top surface. The attachment of target molecules to the top surface leads to a change in surface stress relative to the bottom surface. As a result, the cantilever bends and the amount of deflection can be measured.

In mass-sensing mode, the cantilever – with capture probes attached – vibrates at its resonant frequency. A change in mass of the cantilever causes a slight shift in resonant frequency, which can be measured. The resonant frequency of the unloaded cantilever, f_0, is given by:

$$f_0 = \frac{1}{2\pi}\sqrt{\frac{k}{m}} \tag{3.1}$$

where k is the spring constant of the cantilever and m is the 'effective mass'. (The effective mass is proportional to the actual mass of the cantilever by a factor that depends on geometry.) If we assume that changes to the spring constant on loading are negligible (which is reasonable if the design is right) then the change in mass, Δm, when the target molecules stick can be calculated from:

$$\Delta m = \frac{k}{4\pi^2}\left(\frac{1}{f_1^2} - \frac{1}{f_0^2}\right) \tag{3.2}$$

where f_1 is the new resonant frequency when the target molecules are attached.

Cantilever-based sensors have been used to detect masses as small as those of a large protein or a DNA molecule, but their sensitivity is ultimately limited by interference from Brownian motion in their surroundings, and by the Q-value you met in Block 1. Similar techniques have been developed using bioreactions on the surface of quartz crystals. These can be even more sensitive, capable of detecting masses equivalent to about one hundred water molecules.

SAQ 3.8 (Learning outcome 3.10)

(a) How might you measure the deflection of the cantilever in stress-sensing mode?

(b) A cantilever sensor operating in mass-sensing mode has an initial resonant frequency of 5.00 MHz and a spring constant $k = 0.10\ \mathrm{N\ m^{-1}}$. Calculate the new resonant frequency when a protein of mass 1 fg (1×10^{-15} kg) sticks to the sensor.

3.1.3 Pore-based sensors

By looking more closely at what happens in a cell we can take a different approach to single-molecule detection. Look back at the second type of sensor shown in Figure 3.17. The basic principle is the same as before: a capture probe is used to bind to a target molecule. However, in this case the capture probe lies within a pore, which in turn is supported by a lipid bilayer. You should recognize the lipid bilayer as the basic structure of the membrane that encloses a cell. The cell membrane forms the boundary between a cell and the outside world, or between one cell and its neighbour. So membranes, and in particular the proteins and other chemicals that they support, will have a crucial role in receiving signals from the outside environment and passing them on.

A cell membrane isn't just a phospholipid bilayer. As suggested by Figure 3.9, the membrane also contains proteins that are designed specifically to sit within it, either crossing it or 'floating' on one side or the other. In fact, a typical cell membrane contains about 50% protein by mass, and in some specialized membranes this can rise to as much as 70%. The proteins are loosely bound and can move around fairly freely within the membrane: you can read more about the role of proteins in passing signals across the cell membrane in ▶*Getting the message across: the membrane proteins*◀.

▶Getting the message across: the membrane proteins◀

A critical step in the signalling chain within a cell is the transport of substances across the cell membrane. This can happen in one of three ways: diffusion, transport involving membrane proteins, or transport inside a small 'bubble' of membrane (a vesicle), broken off at one location and then taken up at another. Many small molecules can diffuse across the cell membrane without help. Larger molecules such as amino acids and ions (which may be intrinsically small, but have a high charge density and a surrounding shell of weakly bonded water molecules) cannot diffuse through the membrane; they need help to get across. Help comes in the form of specially adapted membrane proteins.

Membrane proteins fulfil many roles, such as controlling adhesion between neighbouring cells and allowing cells to recognize each other. Here, we are most interested in their role in communication. The membrane proteins can associate with the lipid bilayer in specific ways appropriate to their function, and they provide some particularly fine examples of how the higher-order structure of a protein can be suited to fulfil a particular role. Some of them are situated on one side of the membrane only. They may be embedded in the lipid bilayer, or anchored to it by a protruding lipid chain or by other weak interactions. Others span the membrane one or more times and also have portions of chains extending into both the interior of the cell and the exterior environment, providing a link between them.

For signal reception the membrane-spanning proteins are particularly important. The signals they respond to are often chemical, but specialized proteins also exist that react to light, or to mechanical forces such as touch or sound. Figure 3.19 illustrates some common structural elements. The membrane-spanning portion often takes the form of one or more α-helices (1, 2). Pores can be constructed from a circular array of helices with a hole down the middle. An alternative structure for a membrane-spanning protein is the β-barrel (3), where a β-sheet folds round on itself to form a tube; this is similar in form to the β-can you met earlier, but in this case there is an open channel down the centre.

EXERCISE 3.6

Where would you expect to find concentrations of hydrophobic and hydrophilic side groups in the protein pores shown in Figure 3.19?

Membrane-spanning proteins that include pores have the capacity to pass molecules or ions from one side of the membrane to the other. In some the pores are permanently open, but in others the channels can open and close in response to

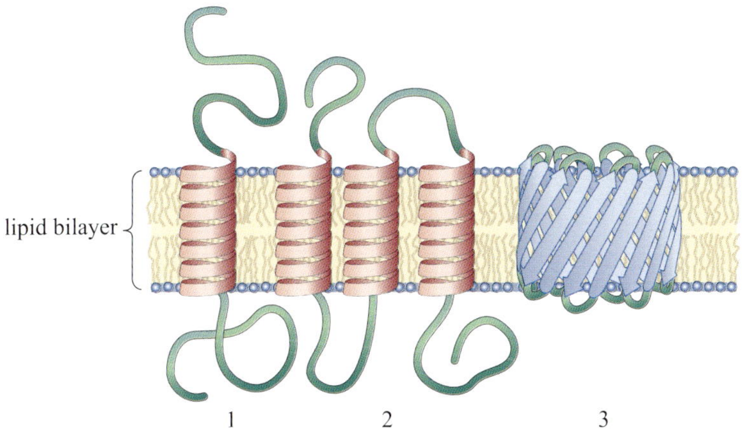

Figure 3.19 Examples of transmembrane protein structures

signals, creating molecular valves. Ion channels that assist the passage of particular species (notably Na^+, K^+ and Ca^{2+}) in response to a signal play a vital role in cells. Shape changes can also form the basis of a motor, converting chemical to mechanical energy (more on this later). Combine a motor with a valve and we have a pump. Valves and pumps can be used to control the chemical composition in different parts of the cell, which in turn can initiate or repress different responses. Other transmembrane proteins work by receiving a signal at the outside boundary of the cell and either releasing, or binding with, a chemical on the inside of the cell in response, often triggering a chain of reactions. The basis of all these changes is a shift in shape within the protein.

You'll see from this that the role of the cell membrane is far more than just containment: it supports a whole industry of dynamic interactions that keep the cell in touch with the world around it.

The pore shown in Figure 3.17(b) is a membrane protein. In the example on which this figure is based, the pores are derived from a naturally occurring protein-based ion channel, where the membrane-spanning portion is a β-barrel. In the cell, the role of the protein is to regulate the flow of ions across the membrane. Outside the cell, it has been found that this flow can be interrupted by molecules known as channel blockers, which bind temporarily to the inside of the barrel. Different versions of the membrane protein can be engineered that bind with different channel blockers.

This provides us with the necessary ingredients for a biosensor. Protein pores are engineered to provide a loose binding site for a specific target molecule. The pores are placed in a planar lipid bilayer, surrounded by salt solution. When a potential is applied across the membrane, ions from the solution flow through it, and the flow of charge constitutes an electric current. When the target molecule binds to the channel it is temporarily blocked, and the current changes as a result. The principle is illustrated in Figure 3.20, with the target molecule represented by a small ball.

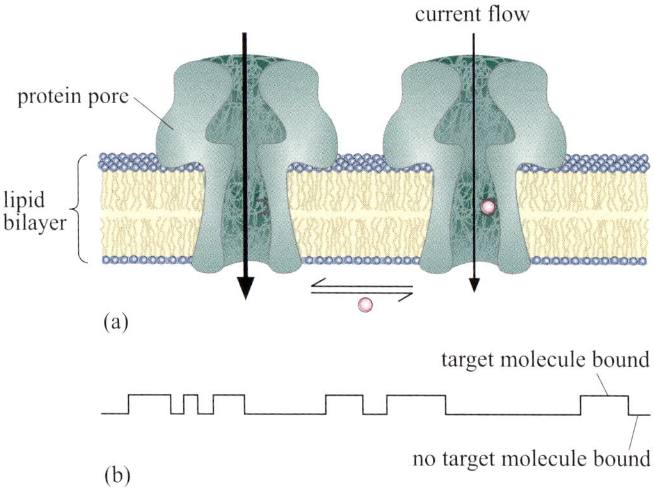

Figure 3.20 (a) An engineered pore in a lipid bilayer, used as a sensor; (b) a typical trace of current versus time. The duration of the pulses is related to the nature of the target molecules, and the frequency to their concentration

Here, the chemical signal from the target molecule has been transduced to an electrical signal. Since only one molecule is needed to block a pore, the sensor can potentially detect single molecules, while for higher populations the frequency of blocking can be related to the concentration of target molecules in the solution. If the pore is sensitive to more than one type of target molecule, then the duration and the amplitude of the signal can be used to distinguish between them.

In common with other types of sensor, many different engineered pores could be combined on a single device to enable several different tests to be carried out simultaneously. A two-dimensional grid containing multiple sensor elements is known as a *microarray*.

SAQ 3.9 (Learning outcome 3.10)

Explain the following features of the pore-based sensor.

(a) The pores are placed in a planar lipid bilayer, surrounded by salt solution. What is the role of the lipid bilayer? Why is the salt solution necessary?

(b) Protein pores provide a loose binding site for a specific target molecule. How might *loose* binding be achieved?

(c) When the target molecule binds to the channel it is temporarily blocked. Why, having attached itself to the binding site, does the target molecule become detached, and why does the period of attachment vary, as indicated by the trace of current versus time?

3.1.4 Signal transduction and amplification in cells

In the two biosensors considered above, a chemical signal has been transduced into either a mechanical response or an electrical signal, both of which can be readily amplified. Other approaches are also being developed: another common method is to label the substance of interest with a visible marker, such as the green fluorescent protein you met earlier, so that its arrival can be detected by optical means. For instance, the presence of a particular bacterium might be detected by exposing it to suitable antibodies, labelled with a fluorescent marker to make them visible. By seeing where the antibodies attach, the distribution of the bacteria is revealed. Within a cell, however, the methods used for the transduction and amplification of signals are, like the initial reception of the signal, largely chemical.

In the simplest case, a signal reaching the outer surface of the membrane surrounding a cell may be able to diffuse directly through the membrane and bind to receptors within the cell, initiating a response of some kind. One common response is to activate a particular section of a DNA molecule and set off the process by which a specific protein is made.

For molecules that can't directly cross the cell membrane a receptor protein on the cell surface is required. These proteins may work in conjunction with ion channels, opening or closing a valve to control the flow of a particular species into or out of the cell.

Other receptors may be linked to proteins within the cell that act as enzymes (biological catalysts that promote certain chemical reactions), or a class of substances known as G-proteins. In this case, binding of the external signalling molecule activates the receptor, which transduces the signal across the membrane by releasing a second molecule inside the cell. This in turn activates the enzyme or the G-protein within the cell and sets off a complex chain of chemical signalling events. G-proteins are particularly common signalling molecules and play an important role in all our senses. Their importance in our bodies also makes them a prime target for drugs and toxins, both natural and synthetic. The Nobel Prize for Medicine was awarded in 1994 to the scientists Alfred G. Gilman and Martin Rodbell, who worked out how G-proteins operate.

Signalling proteins may have multiple binding sites, allowing them to function as integrating devices that switch on the response only when the correct combination of inputs has been received. The ultimate response of a particular cell to a particular signal will depend on which proteins are present, which in turn depends on what type of cell it is. As we learn more about how proteins function inside the cell, so we can develop methods to harness this power outside the cell. The potential for designing sensitive, selective and intelligent sensors is immense.

SAQ 3.10 (Learning outcome 3.5)

(a) What properties would a signalling molecule need in order to pass through a cell membrane?

(b) Describe, in terms of ligands and binding sites, how a signalling molecule would interact with a protein on the surface of the cell.

EXERCISE 3.7

Suggest an example of an engineering application of a biosensor.

3.2 Signal transmission

3.2.1 Chemical and electrical signalling

Where many cells are working together within an organism, methods are needed for passing signals between them. The human body provides a good example: our bodies have developed two main systems for transferring signals between cells at a distance, one relying on chemical methods and one on electrical transmission. Both methods are illustrated in Figure 3.21.

Figure 3.21(a) and Figure 3.21(b) illustrate chemical signalling. For short distances the signalling molecules simply diffuse from one cell to another. For longer-range communication, chemical-signalling molecules (hormones) released by specialized cells are carried around the body in the blood. I expect you have come across a few examples of hormones, as they play very important roles in our lives: adrenaline, oestrogen and insulin are just a few that spring to mind. When the signalling molecules reach their destination they either diffuse across the membrane (as in Figure 3.21(b)) or are picked up by receptors on the surface of the target cells (Figure 3.21(a)), as described in the previous section. Whether the cell receives a signal or not depends on whether it has the right receptors available.

In contrast, for rapid signalling to specific target cells, the nervous system provides a much faster and more directed method for sending signals in the form of electrical impulses, as illustrated in Figure 3.21(c). It is connected up rather like a sophisticated

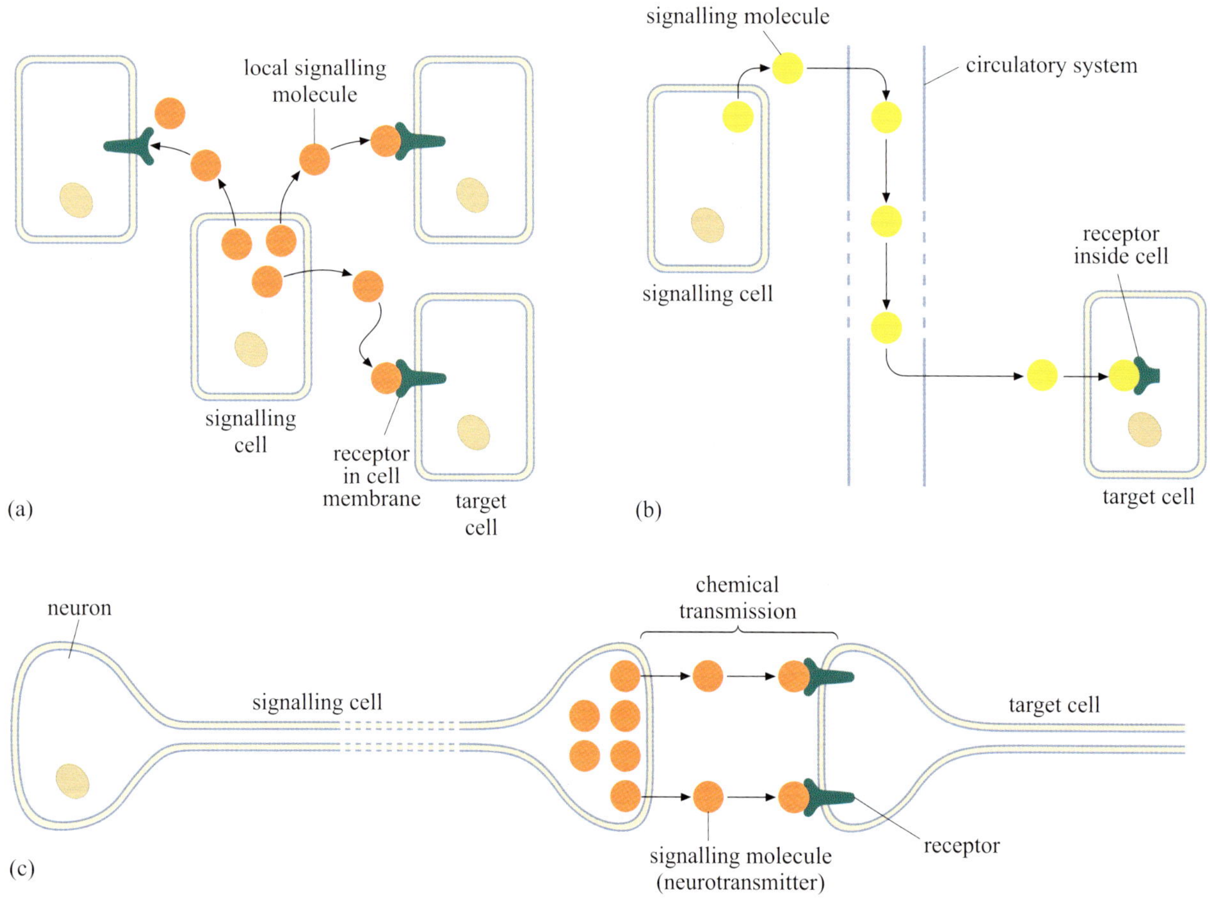

Figure 3.21 Examples of how signals may be transmitted between cells: (a) chemical signals are released and diffuse to nearby cells; (b) for signalling to more distant cells, chemical messengers can be carried around the body in the blood; (c) electrical signalling by specialized nerve cells

computer network, with the brain as the hub, receiving and processing signals carried by electrical impulses from all over the body. These two rather different communication systems work together to maintain the right conditions for life.

The method used for electrical signalling in the body is very different from what we are used to in electrical circuits based on the flow of electrons down wires, and much slower too. The speed with which an electrical signal can pass along a nerve varies from a maximum of around 120 m s^{-1} for muscular signals to a positively sluggish 1 m s^{-1} for non-urgent messages, such as mild pain. As you read this, the message travels at about 20–30 m s^{-1} to your brain. Even the highest transmission speeds are several million times slower than an electrical signal travelling down a wire. So why should we be interested in how the nervous system works?

Many exciting developments in bioengineering, involving the replacement of defective body parts, need to interface with the nervous system to be effective. Artificial analogues of both the retina, which detects light in our eyes, and the cochlea, which detects vibrations in our ears, have been developed, but unless they can send appropriate signals that the brain can interpret they are of little value. At the other extreme, an artificial limb that could understand the electrical signals sent from the brain to control movement would be a fine achievement. It is important, therefore, to understand the principles of electrical communication within the body.

3.2.2 Electrical transmission within cells

The nervous system relies on extremely specialized cells called neurons: an example is shown in Figure 3.22. The main cell body is surrounded by many highly branched protrusions called dendrites, which provide connections with neighbouring cells. Some dendrites are very long – up to about a metre in the spinal cord of a human – and are called axons. Information from the network of dendrites and the cell body is transmitted along the axon in the form of electrical impulses. At its end, the axon is also highly branched, ending in terminals called synapses, which pass the signal on to the next cell. The extreme degree of branching, both in the dendrites and at the end of the axon, means that a single neuron can interact with a very large number of other neurons. Each neuron makes around 1000 links – as many as 100 000 in some cases – so that highly complex networks can be constructed. The human brain is estimated to contain of the order of 10^{12} neurons, so the potential number of interconnections is enormous; vastly more than that found in the integrated circuits of a computer.

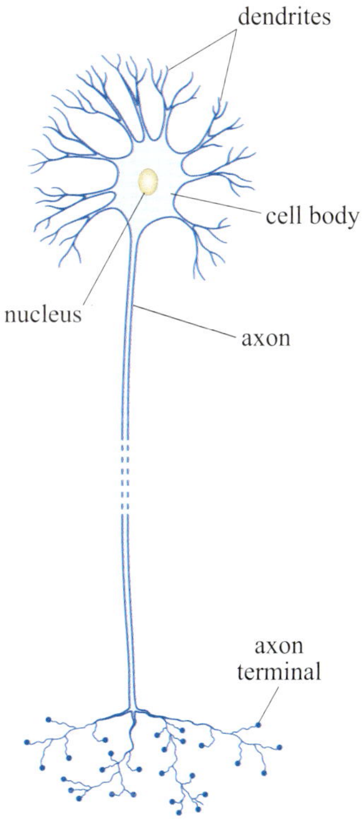

Figure 3.22 Schematic representation of a neuron. The dotted lines indicate that the axon is often very long in comparison to the cell body

It is tempting to think of the axon as akin to a wire in an electrical circuit, but apart from the common feature that both pass charge from one point to another, the mode of operation is quite different. Along with many other processes within cells, electrical signalling relies on the fact that the concentration of a particular chemical species may be higher on one side of a cell membrane than on the other. If the species in question is an ion, i.e. if it carries a charge, then a difference in concentration on either side of a membrane also implies a potential difference across it. The creation of a potential difference across a membrane is a useful way of storing energy within a cell: energy that can be released when required by allowing ions to flow down their electrochemical gradient.

Table 3.4 shows some examples of the concentration of different ions inside and outside a typical mammalian cell. Not all the ions present in the cell are included – the apparent deficit in negative charge is made up by more complex ions such as phosphates – but from these representative values you can see that there are often large differences in the concentration of a particular ion on either side of the membrane. These add together to create a net potential across the membrane known as the *membrane potential*.

Table 3.4 Concentrations of ions inside and outside a typical mammalian cell

Ion	*Concentration inside cells/mmol l^{-1}*	*Concentration outside cells/mmol l^{-1}*
Na^+	5–15	145
K^+	140	5
Mg^{2+}	0.5	1.2
Ca^{2+}	10^{-4}	1–2
Cl^-	5–15	110

Overall the imbalance in ions on either side of the membrane ensures that the normal condition is for the inside of the cell to be more negative than the outside: in fact, membrane potentials range from –10 to –100 mV, with a typical value for the 'resting potential' of a mammalian neuron set at –70 mV. Perhaps a useful analogy is to consider the membrane potential as akin to the voltage appearing across a p–n junction due to the diffusion of holes and electrons.

While calcium ions have many important roles in message transmission, it is the differences in sodium and potassium ion concentrations inside and outside the cell that are most crucial for electrical signalling. Maintaining these differences requires constant pumping of ions against a concentration gradient in a process known as active transport. The molecular pump used to achieve this is described in ►***Pumping ions: the cell membrane sodium pump***◄. Considerable energy is consumed in this process, with up to 70% of the energy used by nerve cells going into ion pumping. The energy required for ion pumping is delivered by a carrier molecule known as ATP. In the next section you'll find out more about the source of the energy carried by ATP, but the basic principle of how it operates is described in the input.

►Pumping ions: the cell membrane sodium pump◄

The concentration of potassium ions (K^+) inside a cell is many times higher than that on the outside: there is a similar difference in sodium ion (Na^+) concentration, but in the opposite direction. Both these ionic gradients are maintained by the operation of a sodium pump: a specialized protein that spans the cell membrane and assists the passage of Na^+ and K^+ ions across the lipid bilayer. Because the pump is working against the concentration gradients of both ions, energy input is needed. The energy is supplied by the conversion of ATP to ADP (see Figure 3.23), which gives the sodium pump membrane protein its technical name of Na^+/K^+ ATPase.

Figure 3.24 illustrates how the sodium pump works. The protein forms a channel through the membrane and protrudes into the cell on the inner surface. This surface has a binding site into which three sodium ions can fit (1). Energy is supplied by the conversion of ATP to ADP, during which a phosphate group (P in the diagram) is attached to the protein, carrying with it energy stored in the bond (2). Attachment of the phosphate group induces a change of shape in the protein, as a result of which the Na^+ ions are no longer bound tightly and are released into the channel and out of the cell (3). A different binding site, into which two K^+ ions can fit (4), is now exposed to the outside of the cell. As the K^+ ions attach themselves to the

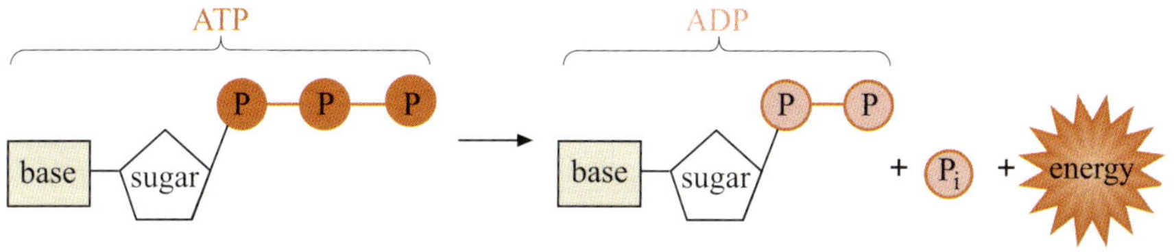

Figure 3.23 Simplified representation of the energy-carrying molecule ATP. Removal of a phosphate ion (labelled P_i) from ATP produces ADP and releases a small amount of energy

protein the phosphate ion (now relieved of its excess energy) becomes detached (5) and the shape of the protein reverts to its original form, releasing the K^+ ions into the inside of the cell (6). So in one cycle of the pump (powered by energy from one molecule of ATP), three Na^+ ions are pumped out of the cell and two K^+ ions are pumped in meaning that there is a net movement of positive charge out of the cell.

An animated version of this figure is provided on the course DVD.

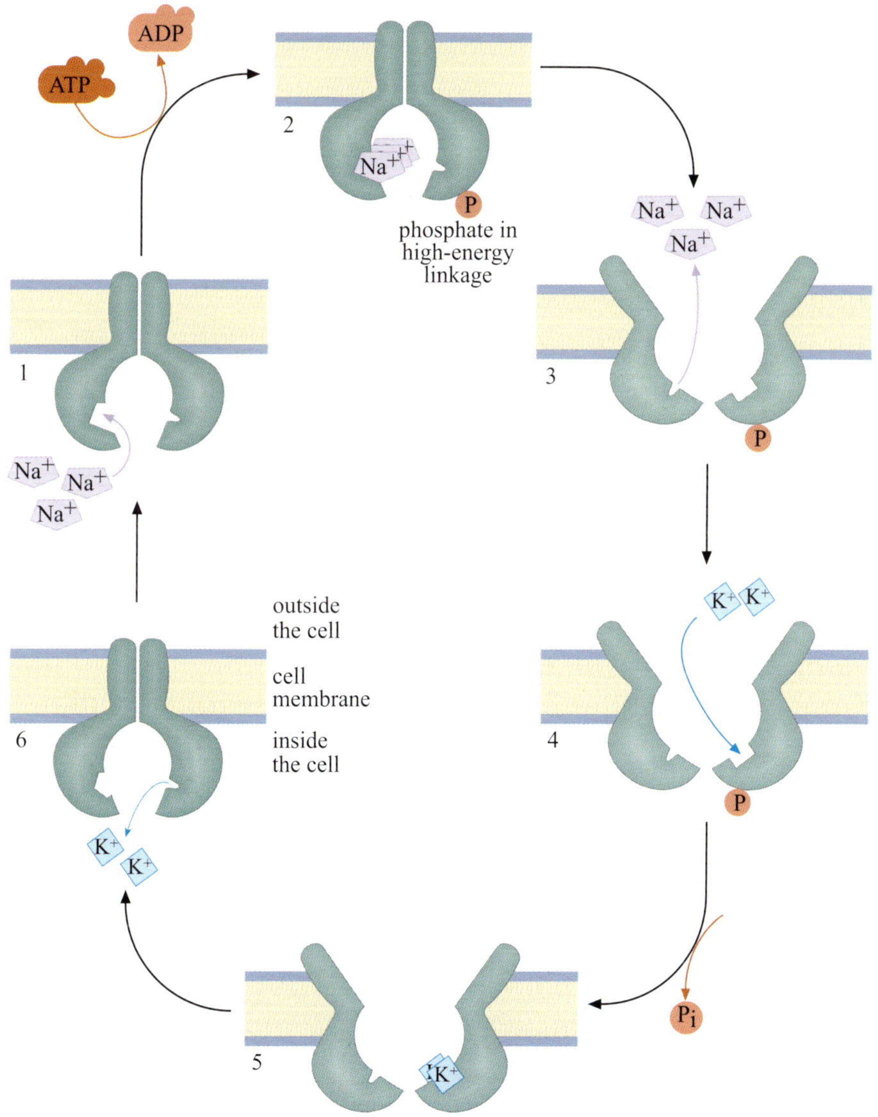

Figure 3.24 The action of the sodium pump that spans the cell membrane

Sodium pumps are present in virtually all animal cell membranes, but are particularly common in nerve cells. Other ATP-powered pumps also exist, and they provide a means of converting chemical energy (from ATP) into stored potential energy in the form of an ion gradient. This stored energy can later be released by allowing ions to travel back down the gradient, releasing power for other essential processes.

You'll come across ATP and its sister molecule ADP a lot in bionanotechnology, as they provide a very convenient way of moving energy around. The A stands for adenosine, but it is more important to remember that TP stands for triphosphate and DP stands for diphosphate. This is the key to understanding how the system works. The backbone of the molecule is a 'base–sugar' combination, which you'll meet again when we look at DNA. You can think of this part of the molecule as a kind of nanoscale handle, onto which the energy-transferring phosphate groups are attached. The handle end of the molecule often acts as a ligand: it has several sites for hydrogen bonding and a small compact shape. Many proteins have binding sites for ATP.

ATP is an unstable molecule. The three phosphate groups have a negative charge, so energy is needed to bring them together. When ATP interacts with a protein the shift in charge distribution results in the detachment of the third phosphate (producing ADP), and energy is released that can be used for other purposes, such as powering the sodium pump shown in Figure 3.24. Biologists often represent a free phosphate ion as P_i and this convention has been used throughout the text.

EXERCISE 3.8

Which species are shown acting as ligands in Figure 3.24?

Action potential is the name given to a large difference in the membrane potential that propagates rapidly along the axon: it is sometimes called the nerve impulse.

Pumping ions across the cell membrane ensures that the membrane potential is maintained. Electrical signals are transmitted along neurons by means of sudden localized changes in membrane potential known as action potentials, or nerve impulses. The process starts with the arrival at the cell membrane of signalling molecules (ligands), which bind to receptors (membrane proteins) at the surface. The binding of a ligand causes a change in conformation, which in this case can cause an ion channel to temporarily open or close.

The transient opening of an ion channel allows specific ions to flow into or out of the cell, causing a small change in membrane potential. Some events may cause an increase in potential, while others may reduce it. Different signals are effectively integrated by the neuron and it is only if the signals reinforce each other to reach a threshold level that a reaction occurs and an action potential is generated.

The threshold level is determined by specialized *voltage-gated* ion channels in the membranes of neurons (and also some muscle cells), which open only when the membrane potential reaches a certain level. The voltage-gated sodium channels crucial to electrical signalling open when the membrane potential of the neuron increases by about 7 mV. Below this level nothing happens. Above this level the channels open and sodium ions flow rapidly into the cell, producing a transient local reversal of the membrane potential. Figure 3.25 shows the flow of ions and the resultant changes in membrane potential as an action potential is passed along the axon of a nerve cell.

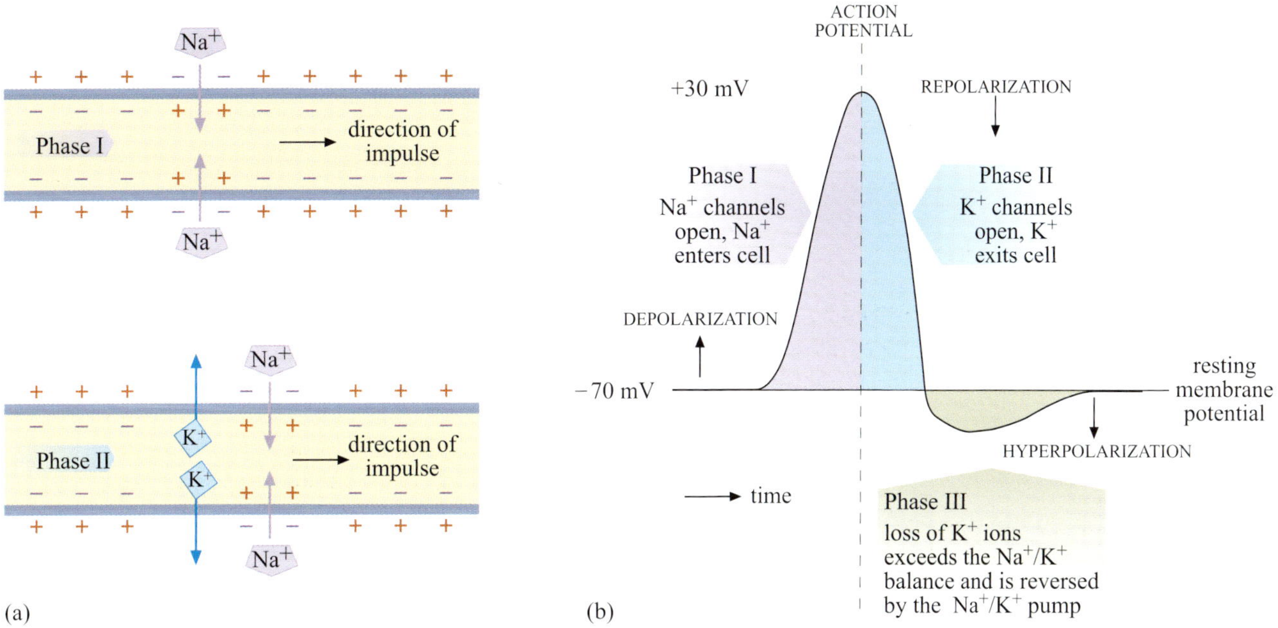

Figure 3.25 (a) Schematic section through an axon showing movement of ions as an action potential moves from left to right; (b) corresponding changes in membrane potential at a single place as the action potential passes

As the voltage-gated sodium channels open, Na^+ ions flood into the cell causing a temporary reversal of the membrane potential (depolarization) at that point: this is labelled phase I in the diagram. The change in membrane potential triggers the opening of a second set of voltage-gated channels, this time allowing K^+ ions to flow out of the cell: phase II in the diagram. As the flow of K^+ ions returns the membrane potential to its original value (repolarization), the voltage-gated Na^+ channels close. As we enter phase III the K^+ channels remain open and the membrane potential briefly falls below its resting potential, before these channels too are closed and the Na^+/K^+ balance is restored by the sodium pump. The final brief reversal of polarization is important as it ensures that the action potential propagates in only one direction: away from the cell body and along the axon. Sodium ions crossing the membrane in phase I will depolarize the membrane immediately in front of them, thus opening new ion channels and passing the signal in that direction, but hyperpolarization prevents the signal returning to where it has just been.

The potassium ion channel is a beautifully elegant example of molecular design for a specific purpose. Figure 3.26 shows a section through the molecule in the membrane. It has a channel down the centre that is designed to act as a selectivity filter, allowing only K^+ ions to pass. The amino acids that line the centre of the pore create an environment that mimics the experience of a K^+ ion when surrounded by water molecules. At the bottom of the channel is the 'gate' region, which opens and closes the channel in response to signals.

When the action potential signals finally reach the axon terminal they are passed on to other cells via junctions called synapses. There are different ways to pass the signal on. The most straightforward is simply via an open pore connecting two cells that allows the ions to flow through: a simple structure based on β-barrels will do the trick. More commonly a chemical, known as a neurotransmitter, is released at one side of the junction and diffuses across to the next cell, where it triggers the necessary response by combining with a receptor. In all this it is the membrane proteins, with their

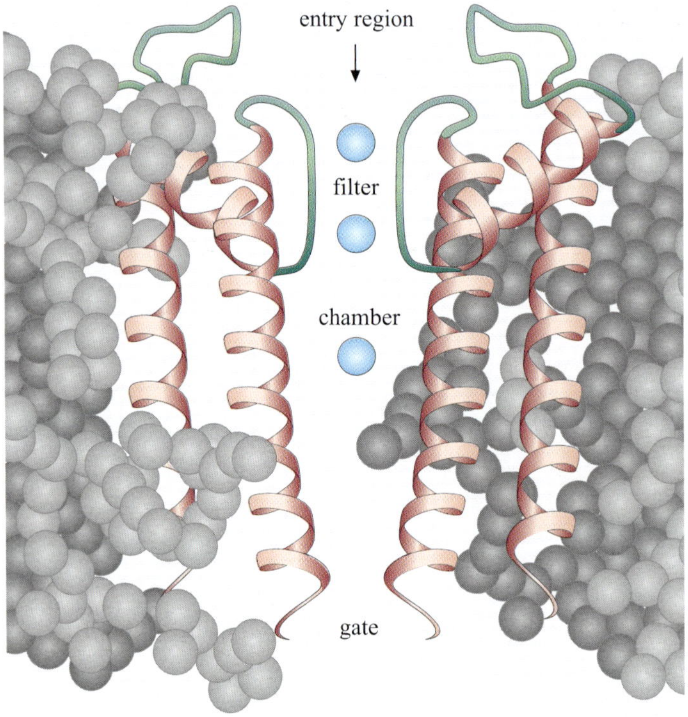

Figure 3.26 The potassium ion channel

specialized binding sites and conformation changes, that hold the key by controlling chemical traffic into and out of the cells.

The system used for electrical transmission in cells is quite different from the one that we are used to in conventional electrical circuits. The action potential is a digitized signal, with fixed amplitude controlled by the trigger voltages of the ion channels. The magnitude of the original stimulus is represented by the frequency of the pulses. These digital signals, together with the highly complex networks created by multiple junctions between neurons, provide the means for programmable computing within our brains.

Perhaps it seems unlikely that this complicated means of passing charge could ever replace the flow of electrons down wires with which we currently design our electrical circuits, but the means to interface the two opens up many possibilities for bioengineering, and voltage-gated ion channels may well find applications in other contexts. One fundamental advantage of biocircuits over conventional manufactured integrated circuits is that they grow in three dimensions, enabling massive connectivity and concentration of devices.

The present focus for the development of nanocircuits is on finding smaller-scale wires, such as electrically conducting polymer molecules, and on building tiny circuit elements, for instance from carbon nanotubes. Where protein molecules may find a role is in the assembly of circuits. By attaching a suitable protein to a nanotube, for example, it can be made to stick to a specific site on a substrate. Self-assembling nanotransistors have been constructed by using protein molecules to attach carbon nanotubes to specific sites on strands of DNA. DNA may itself form the basis of tiny conducting wires – more on this in Section 6.

SAQ 3.11 (Learning outcomes 3.1 and 3.6)

(a) With a diffusion coefficient of ~3×10^{-10} m^2 s^{-1} and a gap between synapses of 100 nm, how long does it take an average neurotransmitter molecule to cross the gap?

(b) The speed of signal transmission along a nerve varies from 1 m s^{-1} to 120 m s^{-1}. Assuming that the difference in speed is entirely due to diffusion between synapses, how many more synapses are there, per cm, in the 'slow' nerve?

SAQ 3.12 (Learning outcome 3.6)

What are the main differences between the transmission of an electrical signal along an axon and the flow of an electric current along a wire?

EXERCISE 3.9

Suggest an example of an engineering application where it is necessary to provide an interface with the nervous system.

4 ENERGY REQUIREMENTS

4.1 Energy in cells

To carry out all their many functions, cells need energy. More importantly, they need energy in a suitable form, and in appropriately sized packages, for the job in hand. Plant cells have a very special capability: they can convert photons of electromagnetic radiation from the sun into stored chemical energy in a process known as photosynthesis. Animals rely on this mechanism too. By eating plants (or, for the non-vegetarian, animals that have eaten plants) they are able both to generate the raw materials for manufacture of new stuff and to release some of the stored chemical energy to power processes within cells.

Of course cells, like everything else, have to obey the laws of thermodynamics. The first law of thermodynamics tells us that energy cannot be created or destroyed but can be converted from one form into another. We will see several examples of energy conversions in this section. The second law of thermodynamics states that for a process or reaction to occur spontaneously, the overall change in entropy must be positive. Entropy is usually interpreted as a measure of the amount of disorder in a system, so an alternative statement of the second law is that the amount of disorder in the universe is constantly increasing – as any parent of small children will know!

Joking apart, it's not immediately obvious how a complex organism like a cell can obey the second law of thermodynamics. There can be few systems more ordered (i.e. with lower entropy) than, say, a globular protein with a single unique folding pattern, yet these and many other complex substances are manufactured within cells. In fact, the stable folding patterns adopted by proteins are the result of a delicate balance between the favourable interactions that lower their energy and the thermodynamically unfavourable decrease in entropy that results. The fact that proteins occasionally fold incorrectly is an indication of just how fine this balance can be, and the difficulty of accurately predicting the energetic balance is another reason why attempting to design new proteins is so difficult. It is also important to remember that cells are 'open' systems and can exchange matter and energy with the external environment, so that an increase in order within the cell can be balanced by a greater decrease in order elsewhere.

Cells carry out processes that require energy by coupling them to energy-releasing processes, so that one can feed off the other. These energy transfers can be brokered in one of two ways:

- using short-lived energy-transfer molecules such as ATP
- via electrochemical energy stored in concentration gradients.

You saw examples of both of these in Section 3. In this section, I will explain how energy from fuel (in animals) or sunlight (in plants) can be converted into these two more useable forms, and how one can be used to generate the other, producing movement in the process.

4.2 Fuel for life: splitting sugar

Animals run on two main types of fuel: fats and carbohydrates. Fats are a type of lipid, while carbohydrates include sugars like glucose and sucrose, and another class of biopolymer – the polysaccharides – formed by linking sugar molecules together. I won't go into details of their structure but will use the sugar glucose, shown in Figure 3.27, as an example of a fuel molecule.

Glucose has the chemical formula $C_6H_{12}O_6$ and is a ring-shaped molecule. It is readily soluble in water, so provides a convenient energy source for a cell. You can get some idea of the energy stored in a sugar by setting fire to some, though I wouldn't recommend trying this at home! Burning fuel releases large amounts of stored energy all at once: excellent for generating heat but quite inappropriate for the delicate nanomachinery that operates within cells. Instead, our food is broken down gradually, or digested, in a sequence of linked chemical reactions. In the course of these reactions intermediate products are formed, some of which may be used in the synthesis of new material, and energy is released in small, controlled quantities. The other essential ingredient, of course, is oxygen, obtained from the air and delivered to cells in the bloodstream.

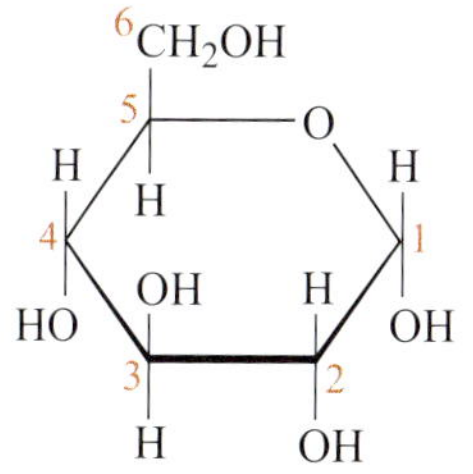

Figure 3.27 The chemical structure of glucose: there are carbon atoms at positions 1–5

Few chemical reactions happen spontaneously – even reactions that ultimately release a lot of energy usually need a 'kick' to get started. This initial energy requirement is known as the activation energy; you may also hear it referred to as an energy barrier. A great many industrial processes use catalysts to speed up reactions and they are important in cells too. Catalysts lower the activation energy for a reaction, so that more molecules have sufficient energy to get over the barrier at any given temperature, but the catalysts are not themselves used up in the reaction. They increase efficiency, and they often have environmental benefits too. Your car probably has a catalytic converter to reduce harmful emissions into the atmosphere.

In the relatively cool conditions of a living cell the reactions by which glucose is broken down would occur much too slowly without the help of catalysts, and once again it is proteins that fulfil this role. Proteins that act primarily as catalysts are called enzymes. You can see some examples of enzymes and read a little more about how they work in Section 13 Proteins in the *Principles* wing. The creation of new catalysts for industrial use, either by 'tweaking' existing enzymes or by using genetic engineering to create new ones, is another important field of bionanotechnology.

The first stage in the breakdown of glucose within cells is known as 'sugar splitting', and as the name suggests it involves splitting the glucose ring (with 6 carbon atoms) into two halves (each with 3 carbon atoms). It is a complex procedure involving ten enzyme-catalysed steps. In energy terms sugar splitting starts as an uphill process, requiring the energy input from two ATP molecules for each molecule of glucose, but in later reactions four ATP molecules are produced: a net gain of two.

EXERCISE 3.10

How is energy supplied to a reaction by ATP? What happens to the ATP in the process?

You saw in Section 3 how ATP can provide energy to a reaction by the loss of a phosphate ion to give ADP; you might like to look back at Figure 3.23 to remind yourself of how this works. The reverse reaction can also occur. A reaction that releases energy can be used to load up ADP with a third phosphate group, storing energy to be used elsewhere. This coupling of energy-releasing and energy-requiring processes is represented in Figure 3.28. The two processes often take place in adjacent binding sites on the surface of a protein.

The amount of energy released by the breakdown of ATP to ADP is just 30.6 kJ mol^{-1}, or 5.1×10^{-20} J per molecule – a tiny quantity, but just the right sort of amount to fuel other processes within the cell. You'll notice a new unit creeping in here: moles (abbreviated to mol).

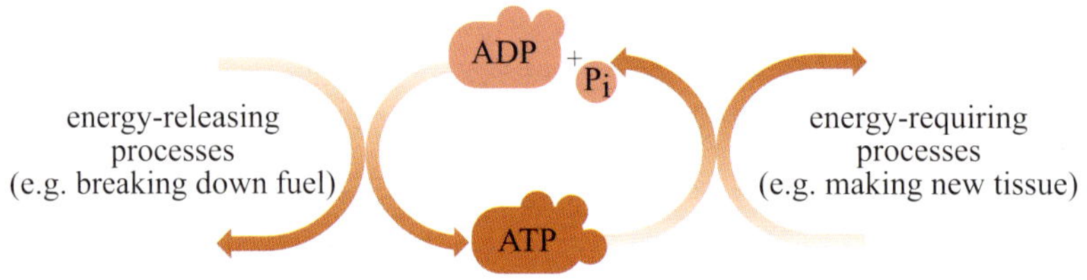

Figure 3.28 Energy-releasing and energy-requiring reactions linked by ATP

Moles are related to the number of atoms or molecules that a sample contains. Because chemists tend to deal in numbers of molecules they usually express energy transfers in kJ mol^{-1}, but we won't be using this unit for calculations in this course. At the level of a single molecule our usual unit of energy, the joule, is embarrassingly large. In fact the electronvolt (eV), which you'll be familiar with from Block 2, would be a far more sensible unit to use here, as you'll see from SAQ 3.13.

SAQ 3.13 (Learning outcome 3.7)

(a) Express the amount of energy (in eV) released during the conversion of ATP to ADP. How does this compare to the sorts of energies involved in the absorption of photons of visible light that you studied in Block 2?

(b) What is the net gain in energy stored in ATP when one glucose molecule is split into two halves?

ATP is often referred to as energy currency for cells. In its loaded-up form it acts as a short-term energy store, releasing energy when its cargo is removed. It can be recycled over and over again, rather like a rechargeable battery, but unlike a battery it is short-lived and needs to be renewed constantly. Sugar splitting is one way in which this is achieved.

4.3 Mitochondrion: the power house of the cell

The final stages in the breakdown of glucose take place in a specialized compartment within the cell. Look back at the animal cell shown in Figure 3.1(a) and you'll see a structure reminiscent of the snake on the badge of an Alfa Romeo, labelled 'mitochondrion'. This is the centre of power production in the cell. The breakdown of glucose is completed here, and there are facilities for storing energy and for making it available, in the convenient form of ATP, when required. The 'snake' is a highly folded membrane, which lies within an outer membrane. A more realistic representation is shown in Figure 3.29.

Mitochondria are found in both plant and animal cells. They can vary greatly in both size (0.5–10 μm) and number per cell (1–1000+), but they all have basic features similar to those shown. The double-membrane structure is particularly important as it provides the means of storing energy in the form of a membrane potential, in this case generated by pumping protons across the inner membrane. Energy stored in this way can be used to power a rotary motor that, as it turns, generates ATP. But let's start from where we left off, with the 3-carbon compounds produced from sugar splitting.

The 3-carbon compounds produced in the first stage of the breakdown of glucose pass through both membranes to the inner chamber of the mitochondrion. Here, a further series of reactions takes place, with the eventual conversion of all of the carbon to carbon dioxide, which diffuses out of the cell as a waste product. In the course of these

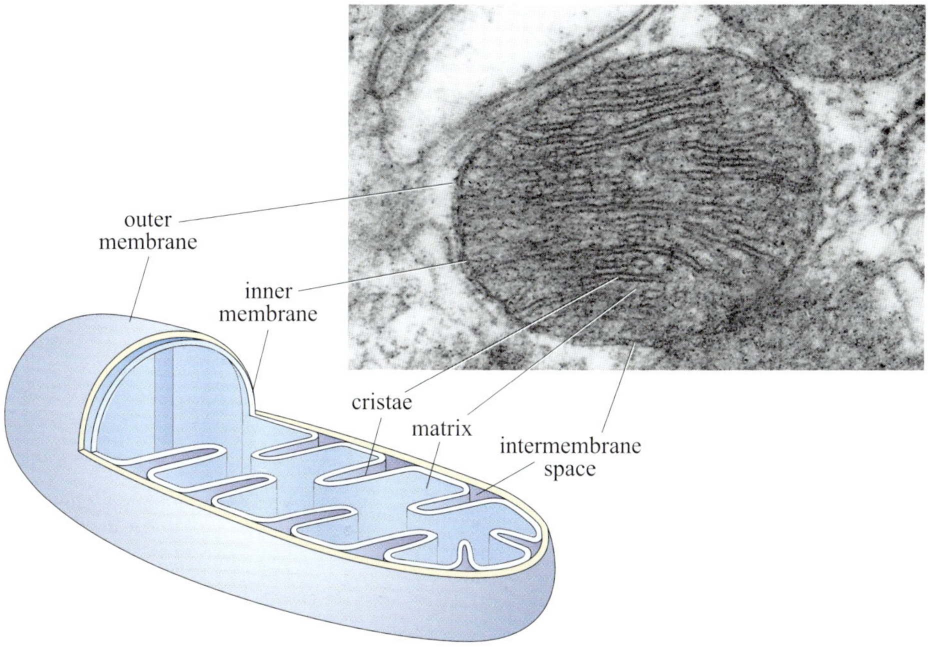

Figure 3.29 Transmission electron micrograph and schematic diagram of a mitochondrion

reactions two more molecules of ATP are generated per glucose molecule; that makes four in total, so far. In addition, hydrogen atoms are produced at various stages in the reaction sequence.

The hydrogen atoms are important for the next stage in the process. They are temporarily attached to a carrier molecule, to be transported to where they are needed. One important carrier molecule is NAD, which can be loaded up with two hydrogen atoms to form NAD.2H.

NAD stands for nicotinamide adenine dinucleotide. There is no need to remember this name.

So far lots of chemistry has taken place, helped along by enzyme catalysts. The carbon ring in glucose has been broken down and carbon dioxide has been formed. Energy released by the reactions has been used to produce a small amount of ATP, and hydrogen is being held in the convenient form of NAD.2H. From a nanoengineering point of view, this is where things get really interesting.

4.3.1 Pumping protons: the electron transport chain

In the next stage of glucose processing, oxygen combines with the hydrogen atoms carried by NAD.2H to make water:

$$\mathrm{NAD.2H} + \tfrac{1}{2}\mathrm{O_2} \rightleftharpoons \mathrm{H_2O} + \mathrm{NAD}$$

This is another energy-releasing reaction and introduces us to a new method of handling energy within cells, using a group of compounds known collectively as the electron transport chain to form a tiny electrical circuit to power a set of proton pumps.

In the previous section you saw how a pulse of charge, in the form of ions, can move along a membrane by the successive opening and closing of voltage-gated ion channels. The electron transport chain (ETC) is another example of charge transfer within cells, this time involving single electrons transferring from one place to another and giving up energy in the process.

The five protein-based components that make up the ETC sit in the inner membrane of the mitochondrion, as shown in Figure 3.30. Proteins 1, 3 and 5 are proton pumps, while proteins 2 and 4 are mobile carriers that can shunt electrons from one pump to the next. The hydrogen atoms are removed from NAD.2H at protein 1 and split into their constituent parts: an H^+ ion (a proton) and an electron. The protons pass into solution on the inside of the inner membrane, while the electrons are transferred to protein 2 to be passed along the chain. Finally, at protein 5, the electrons and protons recombine, and react with oxygen to complete the reaction.

An animated version of this figure is provided on the course DVD.

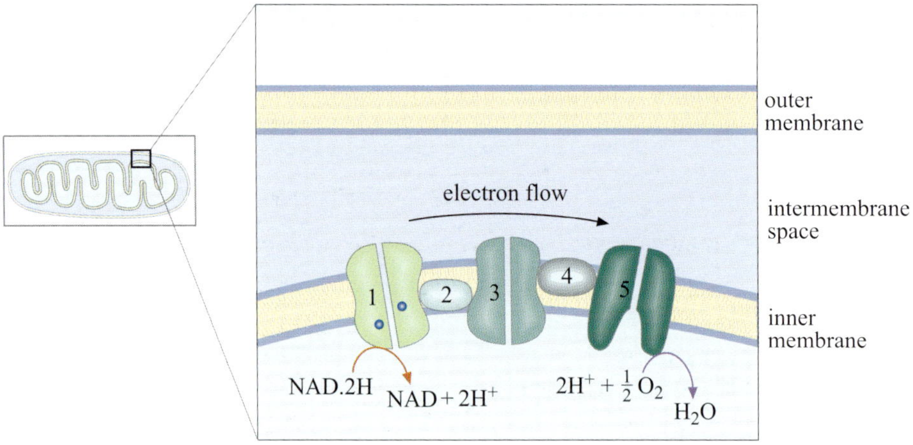

Figure 3.30 The electron transport chain for processing of NAD.2H, showing the direction of electron flow. The relative sizes and shapes of the carriers are known from electron microscopy

Redox potentials and other thermodynamic quantities relevant to cells are measured under neutral conditions (pH 7) to reflect the environment in the cell: the prime in $E^{o\prime}$ serves to differentiate these values from other standard values.

Why are the electrons passed along the chain? Each of the proteins, from 1 to 5, has a stronger affinity for the electrons than the one before, which we can quantify in terms of their redox potential, $E^{o\prime}$. Redox potentials, which you may have met before in the context of corrosion, give a measure of the driving force for a particular electrochemical reaction, relative to a standard reference point. Here, we can regard them as providing a measure of the affinity for electrons of each of the proteins; the higher the value of $E^{o\prime}$, the greater the affinity for electrons. Redox potentials are expressed in volts, so they also allow us to work out the energy available (per unit charge) at each stage in the electron transport chain. The redox potentials associated with electron transport along the chain are shown in Figure 3.31. Note the direction of the *y*-axis scale: it's plotted this way to emphasize that energy is released as electrons pass along the chain.

As the electrons pass along the ETC, they move down a potential gradient and energy is released. As with an electric current, this energy can be used to do useful work: in this case, pumping protons across the membrane. The large drops in potential across proteins 1, 3 and 5 provide the energy for them to transfer a proton from one side of the membrane to the other.

Pumping charge across a membrane to produce a potential gradient provides the cell with a means to store energy. You can think of this as equivalent to a pumped water-storage system for producing electricity, in which surplus current from off-peak periods is used to pump water from a lower to a higher storage reservoir. In this case, electrical energy is converted to gravitational potential energy in the water; this can be turned back to electrical energy when required by letting the water flow back down through water turbines coupled to electricity generators.

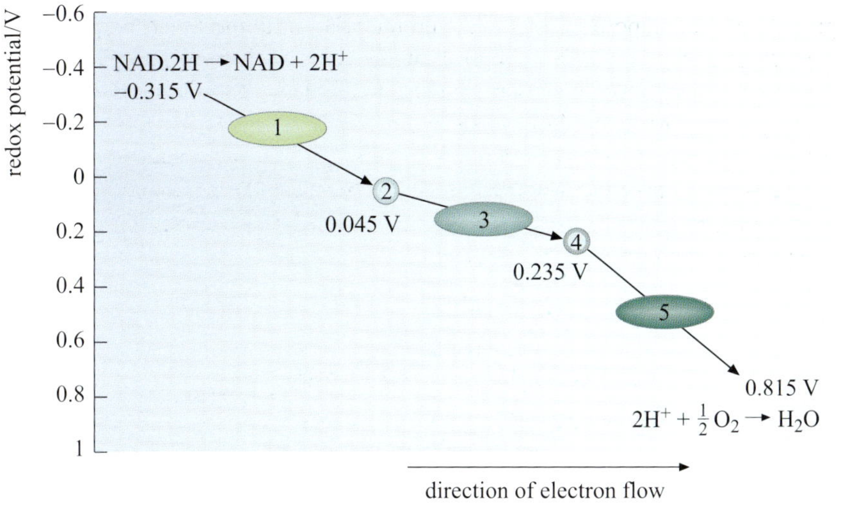

Figure 3.31 Standard redox potentials of electron carriers in the electron transport chain

SAQ 3.14 (Learning outcome 3.6)

(a) In Figure 3.32 the ETC is represented as an electric circuit containing a battery and three pumps. What voltage is provided by the battery and what is its origin? Discuss whether this is a reasonable analogy.

(b) For each molecule of glucose that is broken down, six molecules of water are produced. How many electrons pass along the ETC to produce six molecules of water?

(c) By calculating the drop in potential across proteins 1, 3 and 5, calculate the energy released at each stage as each electron passes down the ETC.

Hint: you will find this calculation considerably easier if you express your answer in eV.

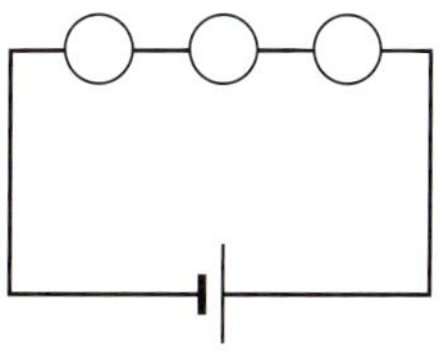

Figure 3.32 Electric circuit representing the ETC

In this analogy, the water is the protons, the pumps are the proton pumps 1, 3 and 5, the higher reservoir is the intermembrane space and the lower reservoir is the inside of the mitochondrion. The equivalent of the turbine is what we'll look at next: a neat little rotary motor and generator called ATP synthase that sits in the membrane alongside the ETC and, as the name suggests, can use the energy stored by proton pumping to synthesize ATP from ADP and phosphate.

4.3.2 Tiny turbines: ATP synthase

Working out the mechanism of ATP synthesis required years of detailed research, and led to the award of the Nobel Prize in Chemistry in 1997 to the scientists involved, Paul D. Boyer and John E. Walker (the prize was shared with Jens C. Skou, who first discovered the cell membrane sodium pump you met in Section 3). The rotary motor has attracted much interest from nanotechnologists, who have demonstrated that it can be used outside the cell to rotate a load.

The basic structure of the enzyme ATP synthase is shown in Figure 3.33. The structure can be split into two main parts, labelled F_0 and F_1 in the diagram. Each of these is a protein-based rotary motor, and they are coupled together via a common stalk down the centre of the structure, labelled γ in the diagram. The lower motor F_0 is embedded

An animated figure showing the different stages of ATP synthesis is provided on the course DVD.

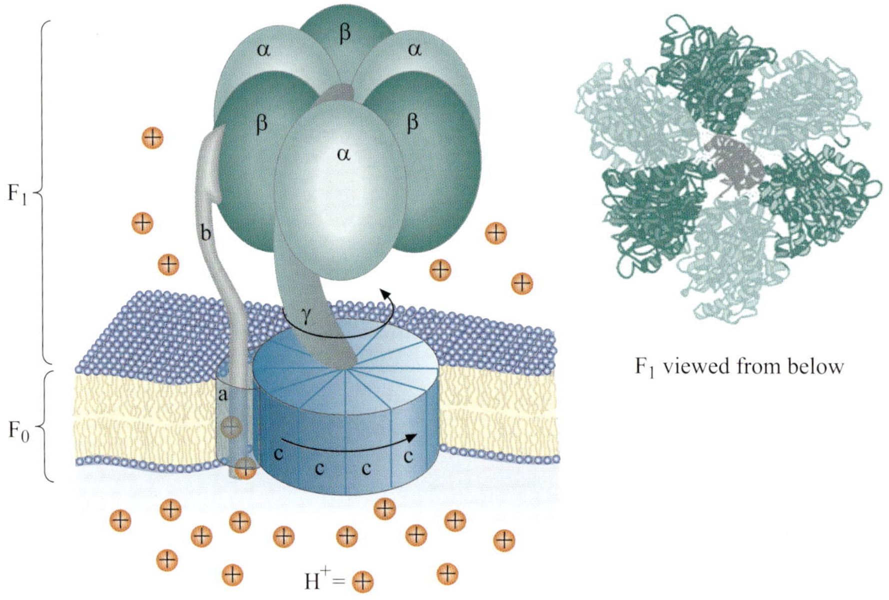

Figure 3.33 The enzyme ATP synthase

in the lipid membrane and driven by the electrochemical energy stored in the proton gradient. It is composed of a moving part – a ring of protein molecules forming a rotor, labelled c in the diagram – and a stationary part – the stator, labelled b. The stator is tethered to the membrane by a third protein, labelled a, which sits next to the ring and guides the flow of protons across the membrane.

A proton flowing through the ring causes it to rotate: a possible model for how this works is illustrated in Figure 3.34. Negatively charged binding sites at the section of the ring adjacent to protein a prevent it from rotating. Binding of a proton neutralizes this charge and weakens the electrostatic interactions that kept the ring tethered in the membrane. Once this has happened Brownian motion is always present to push things around at the molecular level. The ring rotates, eventually depositing the proton on the other side of the membrane. Protein a remains stationary and holds the stator rod b in place so that it constrains the motion of the upper unit F_1.

The upper unit F_1 is like a cylinder, with alternating subunits labelled α and β in Figure 3.33. The view from above shows the conformation of the polypeptide chains; you can probably pick out the familiar helical secondary structures that we have seen in other proteins. It is linked to the lower unit by an asymmetrical crankshaft, labelled γ. The crankshaft is fixed to the ring of c units in F_0, so it rotates with the motor. The α and β subunits in the F_1 part, however, can't rotate because they are locked in a fixed position by the stator rod, b.

Because the axle is asymmetrical it causes distortions in the shape of the β subunits as it rotates. As we have seen before, changes in the shape of a protein can promote a chemical reaction: in this case the production of ATP. The changes that occur as the crankshaft rotates are illustrated in Figure 3.35, which shows the ATP binding site on each of the β subunits. Binding sites are successively opened, allowing ADP and a phosphate ion to enter and combine together to form ATP, and then closed, releasing ATP into the cell. One complete rotation (powered by the passage of one proton across the membrane) produces three molecules of ATP.

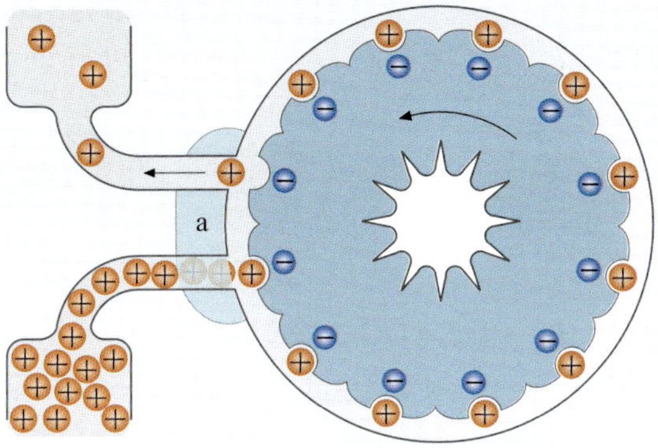

Figure 3.34 The F_0 rotor of ATP synthase binds with protons from one side of the membrane and carries them to the other side as it rotates

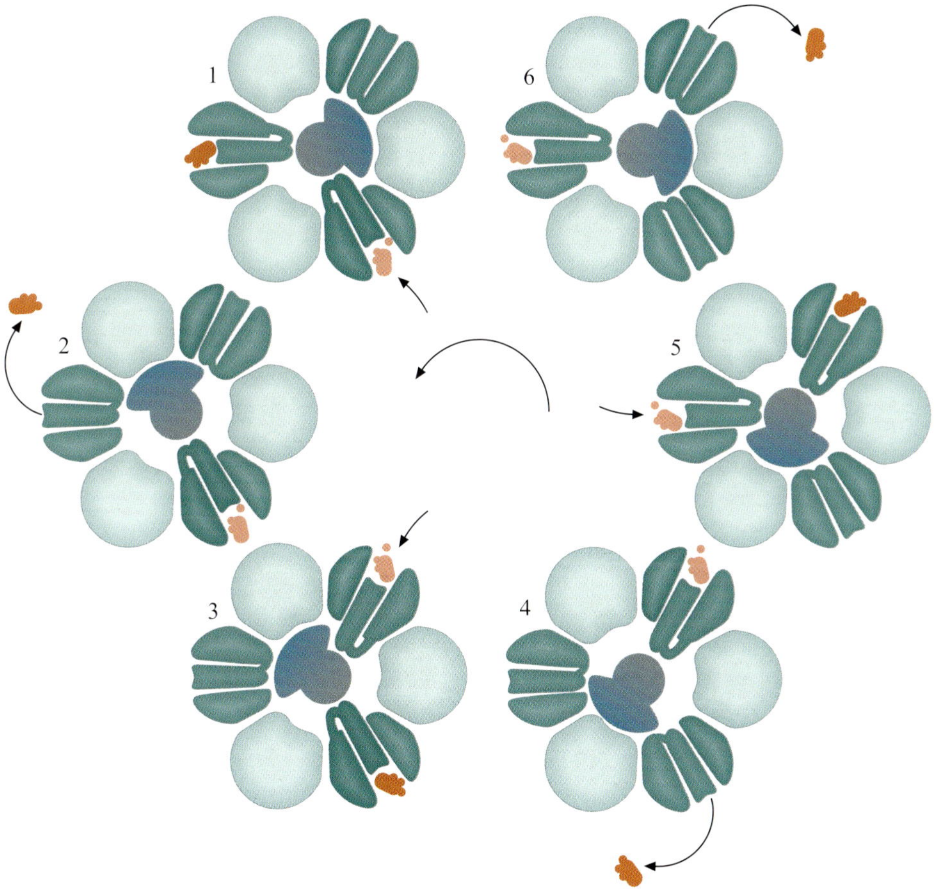

Figure 3.35 The rotary cycle of ATP synthase

I have described here how ATP synthase works as a generator, producing energy-rich ATP for cells to use to power other processes. In this role it is crucial to life. What are the energy transformations occurring when ATP synthase operates in this way? We started with a potential difference across the membrane due to the difference in proton concentration on either side, so we are utilizing electrochemical energy to turn the motor unit as protons move back across the membrane. The motor then rotates the crankshaft; you can think of this as the generation of elastic energy, in much the same way as a key is used to tighten a spring in a clockwork toy. As the crankshaft rotates within the F_1 unit, the elastic energy is transferred to the three β subunits, where it is transformed into chemical energy as the reactions take place.

The ATP-synthase motor can also be run in reverse, using ATP as fuel to produce rotary motion, and it is this aspect that has received more attention from nanotechnologists. ATP-synthase motors have been extracted from cells and incorporated into synthetic devices. In one classic experiment, electron-beam lithography was used to construct a surface coated with a series of nickel-coated pillars, each about 100 nm in diameter and 200 nm high. Tiny propellers, just 150 nm in diameter and about a micrometre long, were also fabricated and coated in nickel. Modified versions of ATP synthase were produced, incorporating groups designed to bind the F_1 section to the nickel pillars and the rotating F_0 section to the propellers. The different ingredients were allowed to self-assemble, with the intended configuration illustrated in Figure 3.36. When ATP ~~synthase~~ was added as fuel, a small proportion of the motors started to rotate, providing an impressive demonstration of the potential for using existing designs from nature as the basis for new nanoscale mechanical devices.

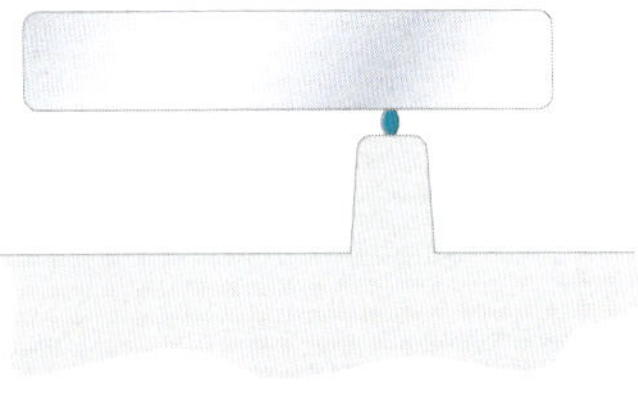

Figure 3.36 An ATP-synthase motor connects a propeller to a post; when ATP is added, the propeller rotates

SAQ 3.15 (Learning outcome 3.5)

(a) With reference to Figure 3.35, explain which parts of this structure are stationary and which parts are moving, and why.

(b) Within the cell, the ATP-synthase generator rotates about a hundred times a second. How much ATP is produced each second? What power does this represent?

SAQ 3.16 (Learning outcome 3.7)

(a) In practice, the combination of chemical reactions, electron transport chain and ATP synthase within a cell results in about 36 ADP to ATP conversions per molecule of glucose. Given that the total amount of energy released when one molecule of glucose is oxidized to carbon dioxide and water is 4.83×10^{-18} J, calculate the efficiency of ATP production in the cell. Comment on your answer.

(b) ATP is short-lived and has to be constantly produced in the body. It has been suggested that an adult turns over a quantity of ATP corresponding to at least half his or her body weight every day. Test the likely validity of this statement, assuming a typical daily energy intake from food of 10 MJ and assuming that all energy transfers within the body are brokered by ATP. The mass of a phosphate group is 95 atomic mass units (1 amu = 1.66×10^{-27} kg).

SAQ 3.17 (Learning outcomes 3.5 and 3.10)

(a) Describe the different stages involved when ATP synthase operates as a rotary motor outside the cell; what are the energy transformations and how are they achieved?

(b) What is the maximum power available from the motor?

4.4 Photosynthesis: harvesting energy from the sun

I started this section by mentioning the key role of plants in energy generation. As well as mitochondria, plant cells have an additional set of energy-processing devices known as chloroplasts. Like the mitochondria, chloroplasts contain many internal lipid membranes, which fold back on themselves to form stacks; these are visible in the electron micrograph in Figure 3.37. These folded membranes provide a very large surface area on which reactions can take place.

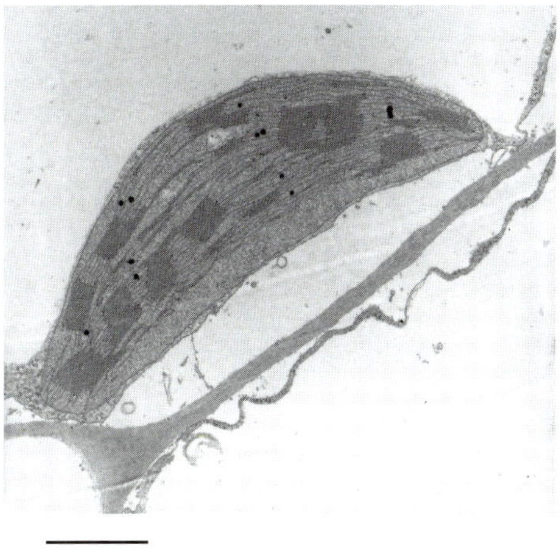

Figure 3.37 Electron micrograph of a chloroplast showing stacked internal membranes

Inside a chloroplast, carbon dioxide and water are combined to produce sugars, such as glucose, and oxygen. The reaction uses solar energy, trapped by the green pigment chlorophyll. While glucose is broken down in the mitochondrion and oxygen consumed, glucose is built up in the chloroplast and oxygen released. This is a hugely important reaction for us because it produces most of the oxygen on the planet, as well as converting solar energy into fuel.

Figure 3.38 shows the processes that take place within the chloroplast – much of this should seem very familiar! The key additions here are the complexes labelled photosystem I and II, which form part of an electron transport chain within the membrane.

Without going into too much detail, what happens here is that photons of light are absorbed at photosystem II and energy from the photons is used to excite loosely bound electrons in molecules such as chlorophyll, creating electron–hole pairs. Chorophyll is effectively a dye; if the electron were allowed to drop back to fill the hole at the lower energy level, it would re-emit the absorbed energy as light of a specific wavelength, producing the characteristic colour of the dye. Electron–hole pairs are also produced by light in a conventional photovoltaic cell (as you saw in Block 2)

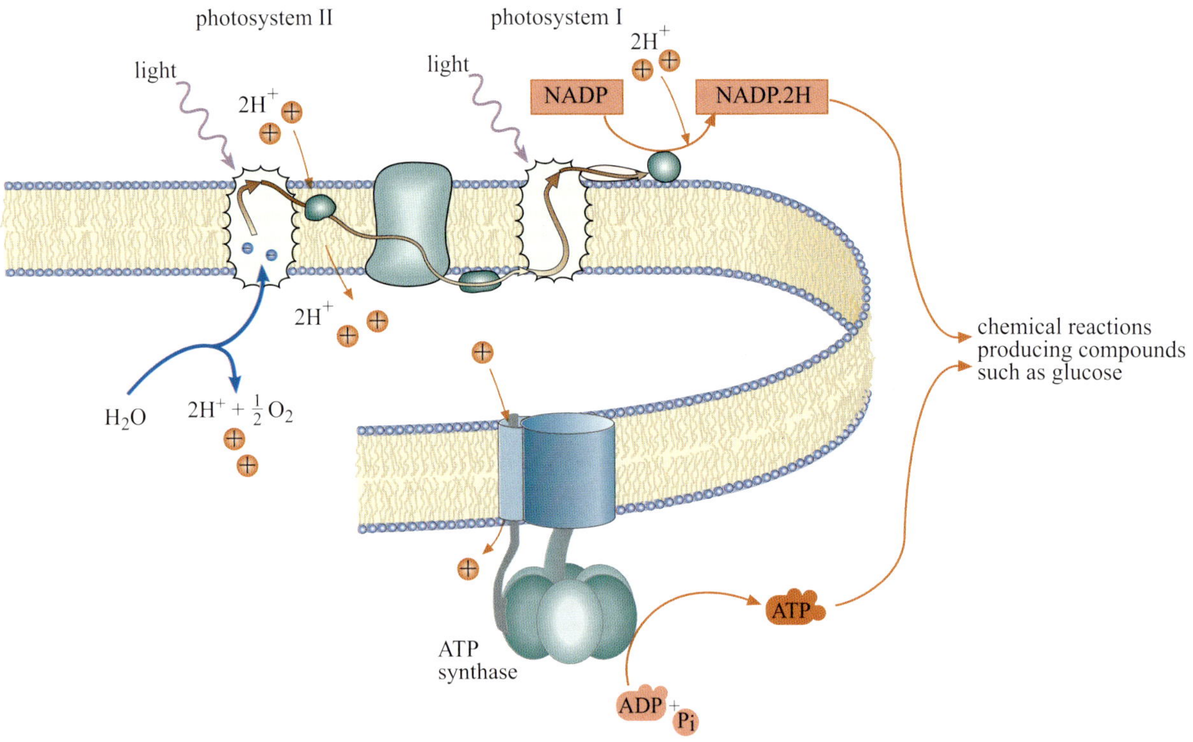

Figure 3.38 Schematic view of the processes that occur within a chloroplast to convert energy from light into ATP

and in that case electric fields can be used to separate the charges and convert the solar energy to electrical energy.

Photosynthesis has two stages, known as the light and dark reactions. Here I have described only the first stage: the light reaction, which generates ATP and H^+ ions. Sugars and carbohydrates are built at the second stage: the dark reaction.

What happens in the chloroplast is rather different. The excited electron is rapidly removed from the vicinity of the hole and passed along an electron transport chain, so it can't drop back down to fill the hole. Instead, the holes left in the chlorophyll molecules are filled by electrons generated by splitting water molecules. Water is not a particularly good electron donor as it takes a lot of energy to split the molecule – more than that delivered by a single photon of light – but there is plenty of it, so cells have adapted to concentrate enough energy in one place to achieve the desired result. The overall reaction at photosystem II is:

$$H_2O + 2\text{ photons} \rightarrow 2H^+ + 2e^- + \tfrac{1}{2}O_2$$

As the electrons pass along the electron transport chain, they provide the power for a mobile carrier to shuttle protons across the membrane (indicated by the brown arrow on Figure 3.38), before being given another boost of energy from a second centre known (for historical reasons) as photosystem I. At the end of the ETC, each electron combines with a proton from the other side of the membrane to form a hydrogen atom; this is then picked up by the carrier molecule NADP and taken off to participate in sugar-forming reactions. The proton gradient created by the reactions at either end of the ETC, and by the proton pump within it, fuels an ATP-synthase rotary motor to generate ATP from ADP and phosphate.

So here we come full circle. We have seen how glucose is broken down in mitochondria to provide energy to fuel cells. In the final stage of photosynthesis energy from the sun, now conveniently packaged as ATP, and hydrogen atoms carried by NADP.2H are used to convert carbon dioxide to, amongst other things, glucose.

SAQ 3.18 (Learning outcome 3.7)

Use Figure 3.38 to answer these questions.

(a) List the processes that contribute to the creation of a proton gradient across the membrane.

(b) What is the energy stored in the proton gradient used for?

(c) What role is played by NADP?

(d) List the energy transformations represented in the figure.

Technologists have already developed silicon-based methods for harvesting energy from the sun and converting it to electricity, but the costs are high. Photosynthesis provides us with the inspiration for developing alternative approaches that could turn out to be much cheaper and that could deliver not just electric current but transportable fuel, such as hydrogen (for large-scale applications) or ATP (for nanomachinery). Many avenues are being explored using the basic principle of photosynthesis: rapid separation of an electron–hole pair created by light, so that the electrons can be collected as electric current before recombination takes place.

Some of these methods use crystalline nanoparticles (titanium dioxide is a popular choice), in combination with a light-sensitive layer, to separate electron–hole pairs and generate a photocurrent. Other approaches are more closely derived from nature and utilize naturally occurring components, sometimes in combination with synthetic molecules, to produce simpler versions of cellular machinery. Artificial photosynthetic membranes based on phospholipid liposomes have been developed and provide one promising line of research.

Energy from sunlight may thus be transformed into an electric current, an electrochemical gradient across a membrane or, by including an ATP-synthase generator in the design, chemical fuel in the form of ATP. While ATP provides an ideal energy source for nanomachines, the generation of electric current or electrochemical gradients is more likely to be used to produce hydrogen from water. Other approaches to producing hydrogen gas directly from cells are also under investigation, all aimed at reducing our dependency on increasingly scarce fossil fuels.

EXERCISE 3.11

Energy transfers within cells rely heavily on the catalytic properties of proteins acting as enzymes. Suggest an engineering application for an enzyme catalyst.

5 TRANSPORTATION

5.1 More molecular motors

In the previous section you saw how ATP is produced in cells by ATP synthase, the ATP generator that can work in reverse as a rotary motor. Within the cell the primary function of ATP synthase is not to produce motion, although it is often this aspect that interests nanotechnologists. Rotary motion on a larger scale is quite rare in nature, but can be seen in some ▶*swimming bacteria*◀.

▶Swimming bacteria◀

Some bacteria move through water using a whip-like tail, called a flagellum, as a kind of propeller to move them along. An example is shown in Figure 3.39. This is based on a rotary motor not dissimilar to the ATP-synthase motor that we examined earlier, powered by the movement of ions down an electrochemical gradient. In this case the cell wall has a double-membrane structure, with a rigid layer in between that holds the stator.

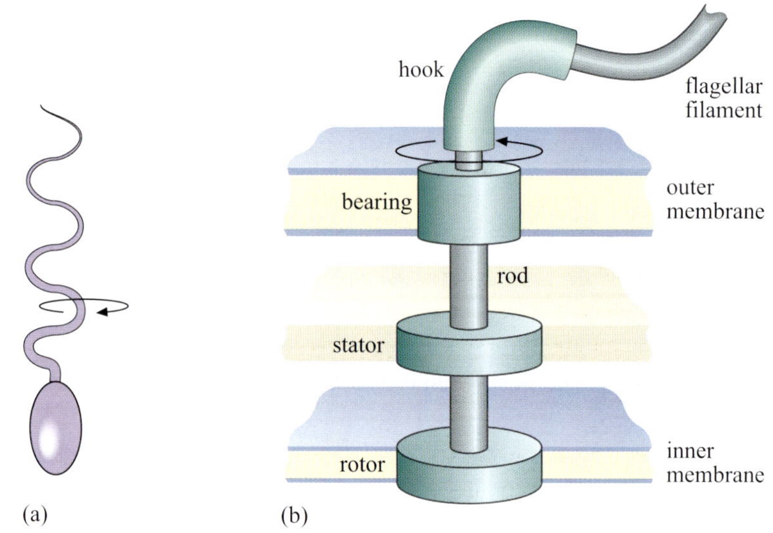

Figure 3.39 (a) Propeller action of bacterial flagellum; (b) example of the structure of the rotary motor that turns the flagellum

It turns out that rotary motion is particularly effective for overcoming the viscous forces that dominate at small scales. Rotations of up to 300 Hz have been observed, and ten filaments acting together can produce speeds of several cell-lengths per second, despite the large resistance due to the viscosity of the surrounding fluid. By changing the direction of rotation, the bacterium can switch from directed forward movement to random tumbling, and in this way its general direction can be controlled.

Linear motors are also found within cells, where they are used for moving things around in a more directed manner than that offered by random diffusion, as well as for larger-scale motion often involving the cooperative action of many cells.

As you know, a motor is a device that converts some kind of energy into mechanical work. Early attempts to make tiny synthetic motors based on miniaturization of conventional designs soon ran into problems of scale. A more effective method for producing motion on the micro or nano scale is to use piezoelectric crystals. These, as you have seen in Block 1, generate very small, controllable increments of movement in response to changes in electrical field.

Stimuli-responsive polymer gels are another class of materials that can change shape, and thereby produce motion, in response to external conditions. These are synthetic polymer networks swollen with solvent in which the shape of the chain is sensitive to factors such as temperature or acidity, so that a change in conditions can induce the chain to stretch out or become more tightly coiled. Small-scale motors are just one potential application of these versatile materials.

Nevertheless, you will probably not be surprised to find that the linear motors within cells are based on proteins – the motor proteins – in which chemical energy is used to produce a physical force that can be translated into movement. An example of a protein-based linear motor at work is described in ▶ ***First robot moved by muscle power*** ◀.

▶ First robot moved by muscle power ◀

A silicon microrobot just half the width of a human hair has begun to crawl around in a Los Angeles lab, using legs powered by the pulsing of living heart muscle. It is the first time muscle tissue has been used to propel a micromachine.

This distinctly futuristic development could lead to muscle-based nerve stimulators that would allow paralysed people to breathe without the help of a ventilator. And NASA, which is funding the research, hopes swarms of crawling 'musclebots' could one day help maintain spacecraft by plugging holes made by micrometeorites.

Whatever the ultimate applications of the technology, no one was more surprised to see the tiny musclebots finally move than Carlos Montemagno, the microengineer whose team is developing them at the University of California, Los Angeles.

He has spent three disappointing years trying, and failing, to harness living muscle tissue to propel a micromachine. But when he and his team looked into their microscopes, they were amazed to see the latest version of their musclebot crawling around.

The device is an arch of silicon 50 micrometres wide. Attached to the underside of the arch, the team has grown a cord of heart muscle fibres (see Figure 3.40). It is the contraction and relaxation of this cardiac tissue that makes the arch bend and stretch to produce the bot's crawling motion. And the muscle is fuelled by a simple glucose nutrient in a Petri dish.

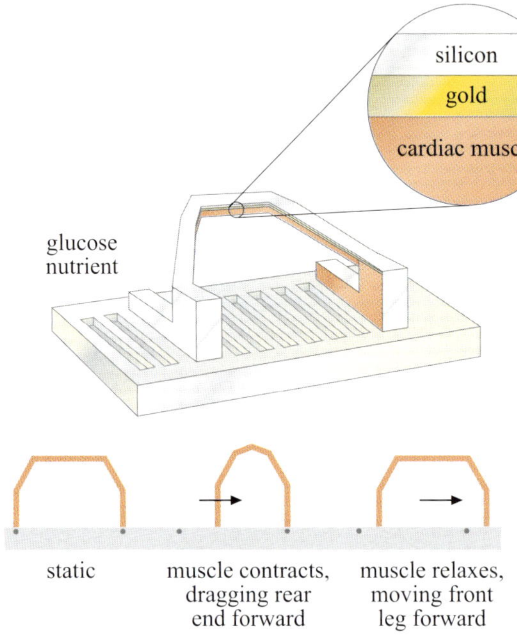

Figure 3.40 Muscle-powered micromachine

Arch-shaped skeleton

The prospect of using living muscle to power microelectromechanical systems (MEMS) is an attractive alternative to micromotors. While motors need electricity, muscles can draw their energy from glucose – perhaps deposited on the surface where the robot will be working.

The UCLA team's breakthrough is to have developed an automated way of anchoring muscle tissue to a substance like silicon. The team carved an arch-shaped skeleton from a wafer of silicon using automated microchip manufacturing equipment, and coated it with an etchable polymer.

They then etched away the coating on the underside of the arch and deposited a gold film there. This acts as an adherent for the muscle cells. To grow the muscle, the skeleton was placed in a Petri dish containing rat cardiac muscle cells in a glucose culture medium. Over three days, the muscle cells grew into muscle fibres that attached themselves to the gold underside, forming a cable of cardiac muscle running the length of the arch.

During this process, the arch was held in place by a restraining beam. When this was removed the musclebot immediately started crawling at speeds up to 40 micrometres per second. The geometry of the musclebot ensures that its flexing pushes it in one direction, rather than simply contracting and relaxing on the spot.

Phrenic nerve

Montemagno now wants to use the technology to help people who have damaged phrenic nerves. These stimulate the diaphragm to make us breathe and damage means patients often need ventilators instead.

Rather than moving the legs of a musclebot, the muscle fibres would flex a piece of piezoelectric material and generate a few millivolts to stimulate the phrenic nerve. Using cells from the patient's own heart would prevent rejection of the implant, and the muscle could be powered by blood glucose.

Montemagno's initial brief from NASA's Institute for Advanced Concepts was to design a muscle-powered micromachine that could seek and repair micrometeorite punctures on spacecraft.

However, he stresses that such applications are several decades away. 'The issue of all of the microbots talking to one another hasn't even been addressed,' he stresses. Or, indeed, how they would be fuelled. Watch out for the sugar-coated space station.

(Source: Ananthaswamy, A. (2004) 'First robot moved by muscle power', *New Scientist*, vol. 181, no. 2436.)

There are three general families of motor proteins – kinesins, dyneins and myosins – although they all act in a similar way and their structures have much in common. The 'musclebots' described in the input use myosin motors, which power our muscles, while the kinesins and dyneins are widely used within cells to transport cargo around. Each type of cell will possess a collection of motors dependent on its particular transport needs.

Protein-based molecular motors use chemical energy from ATP to generate movement. Look back at the information about ATP in Sections 3 and 4 if you need to, to remind yourself how chemical energy is stored in ATP, how ATP interacts with a protein, and what happens when it does.

The motors produce motion by interacting with filaments within the cell. These filaments are further examples of structural proteins. The different types are described in ►*Cellular scaffolding*◄.

►Cellular scaffolding◄

Images of cells, such as the AFM picture shown in Figure 3.41, reveal a great deal of fine, fibrillar structure within the cell. This is the cytoskeleton – a network of filaments that give the cell shape and internal support, and play a crucial role in internal transport and cell movement.

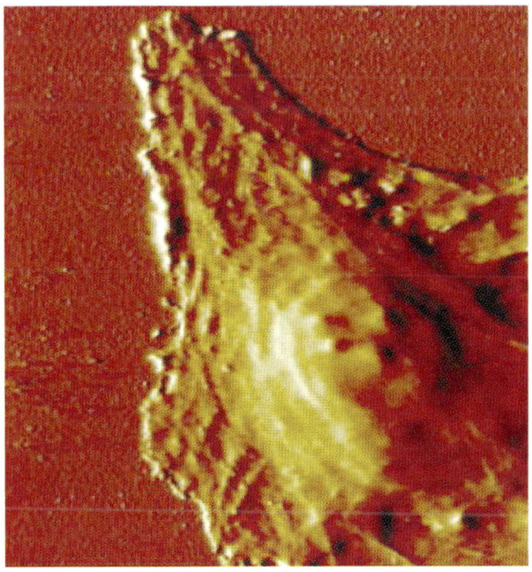

Figure 3.41 AFM image of a living cell

The cytoskeleton is often referred to as the scaffolding of the cell. This is a useful description, yet it hides the fact that the structure is not static, but constantly changing as the filaments are assembled and broken up in response to the needs of the cell.

The cytoskeleton is made up of proteins, arranged into long, filament-like molecules. There are three types of filament, each based on a different protein, and these are illustrated in Figure 3.42.

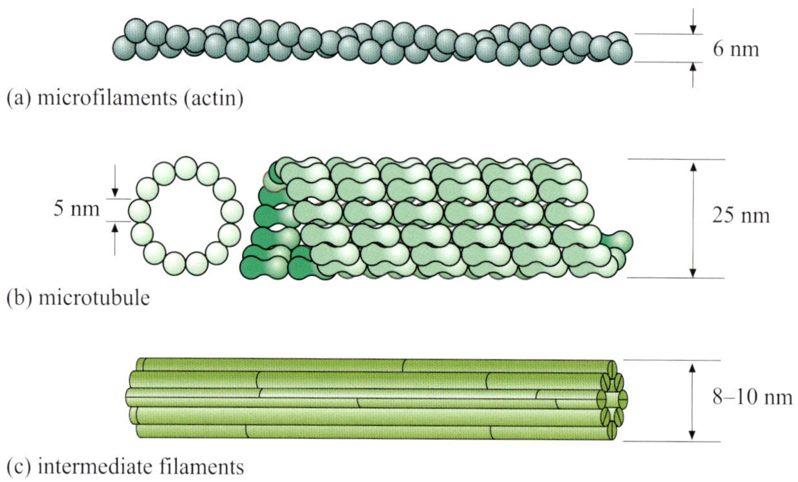

Figure 3.42 Components of the cytoskeleton: (a) microfilaments; (b) microtubules; (c) intermediate filaments

Microfilaments (sometimes called actin filaments) are made from the protein actin and tend to be found around the edges of cells. They are particularly important for muscular movement, as you will see when we look at the myosin motor. Microtubules are larger and are made up from subunits of the protein tubulin: these polymerize to form long filaments, which in turn organize into hollow tubules formed from thirteen parallel filaments. Microtubules grow out from a central point and provide tracks for moving materials around inside cells. Intermediate filaments come in different types, but all have a multi-stranded structure that conveys great strength and therefore has a key role in mechanical support.

SAQ 3.19 (Learning outcome 3.3)

(a) What are the main similarities and the main differences between the structures of collagen and actin?

(b) A single filament of actin is made up of two strings of actin units, each approximately 5 nm in diameter. A straight filament has been found to break under a force of 600 pN. Calculate the tensile strength of actin and compare it to the value of 550 MPa measured for collagen. Comment on your answer.

(c) The force needed to break an actin filament drops rapidly if the filament is flexed, but the same is not true for collagen. Explain this observation.

The cytoskeleton is a dynamic structure that is continually being built up and broken down within the cell, as a result of shape changes brought about by ATP to ADP conversions. Microfilaments provide the tracks for the myosin family of molecular motors, and can even be used to move whole cells around by selective extension and retraction of filaments. Microtubules form the tracks along which the other two families of motor proteins – kinesins and dyneins – travel. Their main role is to move things around inside the cell.

Figure 3.43 illustrates one example of how molecular motors are used to carry cargo. In the nerve cell shown, where transport over relatively large distances is required, signalling molecules packaged in vesicles (small lipid bubbles) are carried from one end of the axon to the other by kinesin and dynein motors moving along microtubule tracks. Because the track is asymmetrical, one direction is likely to be favoured over the other by a particular type of motor. Until recently it was thought that each family of motors travelled in only one direction along a track, but exceptions have recently been found that show this to be untrue.

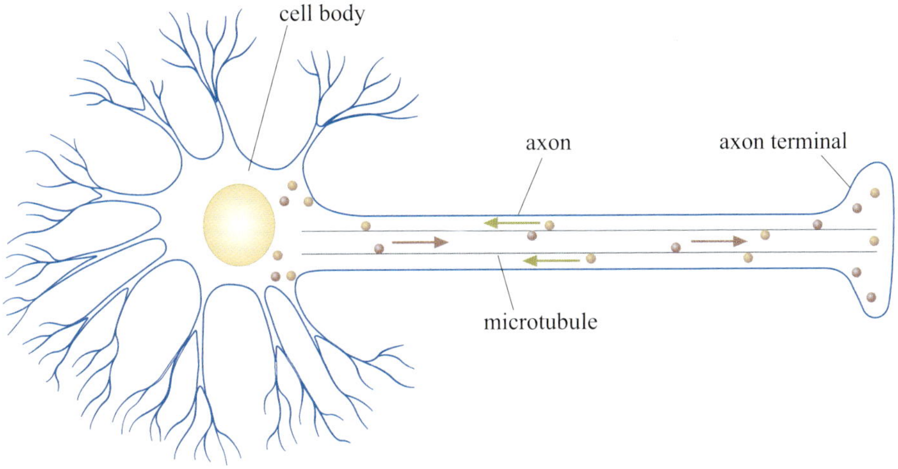

Figure 3.43 Molecular motors carry cargo, packed in lipid vesicles, along the axon of a nerve cell

A common analogy is to think of the microtubules as railway tracks on which the motors ride. However, research has shown that the motion of these motor proteins is actually much more like a walking motion, where adjacent parts of the molecule are alternately attached and detached as the motor travels along. So perhaps imagining a stroll down the road, wearing particularly sticky shoes and carrying a rucksack, provides a better picture of what is going on.

The role of ATP in this process is twofold. Firstly, it provides the energy to power the movement; this is achieved in the usual way by the conversion of ATP to ADP, via a binding site on the surface of the protein. The conversion occurs at the end of the protein that interacts with the track; this end is usually called the 'head' of the motor protein, although 'foot' would be a better description for our walking analogy! In addition, the small change in shape that accompanies the ATP/ADP conversion at the head is amplified to produce a much larger motion elsewhere. This is illustrated in Figure 3.44 for kinesin and Figure 3.45 for myosin. Dynein is more poorly understood.

Both kinesin and myosin are dimers: they consist of two molecular segments twisted together to produce a two-headed structure with a long tail. Let's consider the interaction of kinesin with a microtubule track shown in Figure 3.44.

Kinesin motors use their double-headed structure to 'walk' along a microtubule track, carrying cargo attached to the tail. Each of the head units has an ATP binding site; this may contain ATP, ADP plus a free phosphate ion, or ADP. As the cycle starts (1) one

An animated version of this figure is provided on the course DVD.

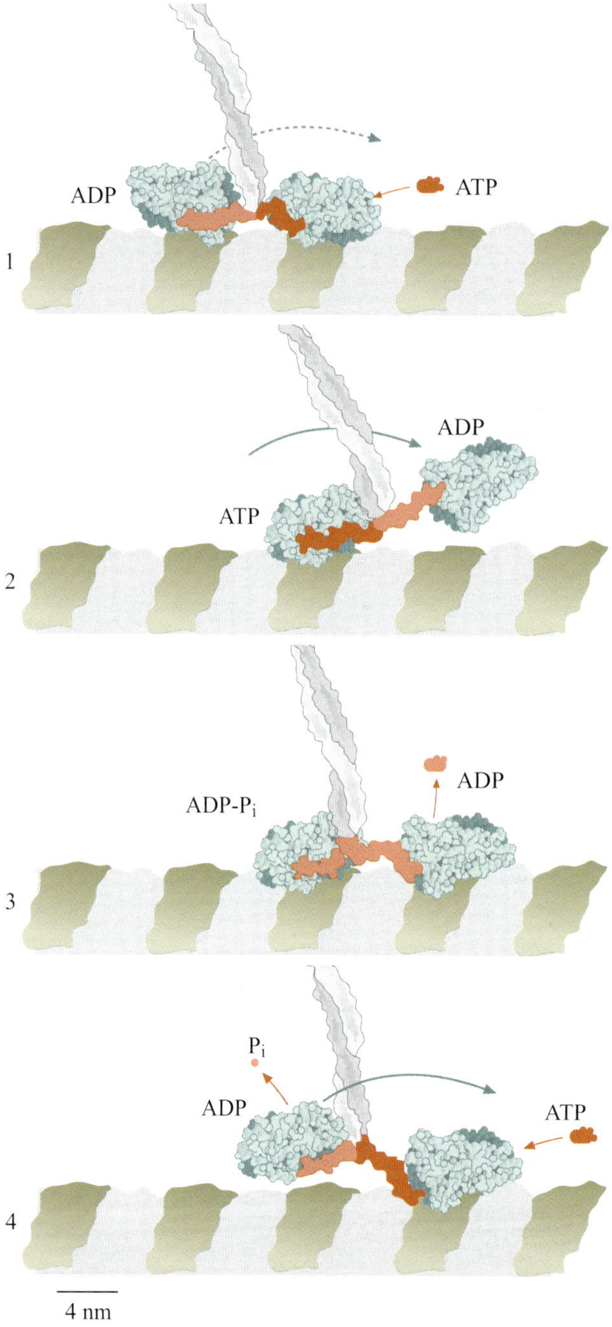

Figure 3.44 ATP-powered movement of kinesin motor protein

site is filled with ADP and a molecule of ATP is binding to the other. The binding of ATP causes a dramatic conformational change in which one section of the protein chain is pulled tightly into a groove on another part of the molecule. The effect of this is to pull the lagging subunit off the microtubule and forward into the next position (2). This provides the power stroke for the motor. As the unit binds in its new position ADP is released from the forward end and the ATP, now at the back, breaks down into ADP plus phosphate (3). Finally, new ATP binds at the front and the phosphate leaves the binding site at the back, restoring the original conformation (4). The cycle is now ready to start again to produce the next step. The kinesin motor stays in contact with the track at all points in the cycle.

An animated version of this figure is provided on the course DVD.

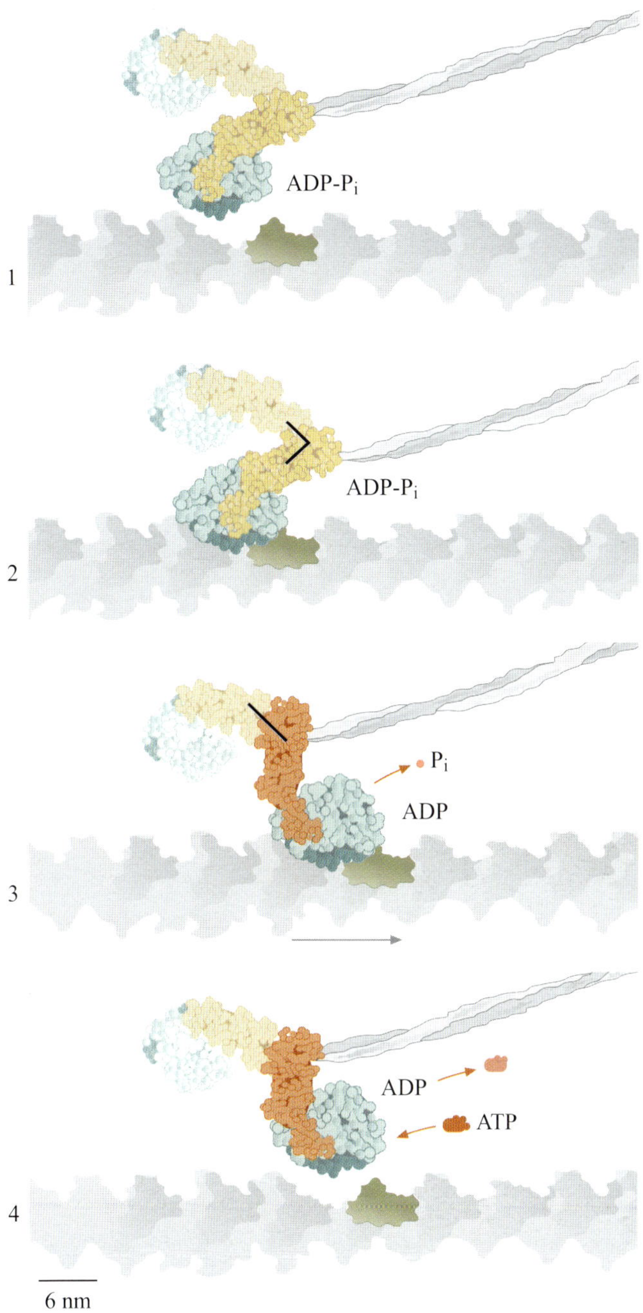

Figure 3.45 ATP-powered movement of myosin motor protein. Yellow and red have been used to differentiate between the two different positions of the lever arm

Now consider the interaction of the myosin motor with an actin track, as shown in Figure 3.45.

Like kinesin, each head of the myosin molecule has a binding site that can hold either ATP, ADP and a phosphate ion, or ADP, and the way in which it interacts with the actin track depends on the contents of the binding site. With ATP in the binding site, myosin does not bind strongly to actin. However, when ATP breaks up into ADP plus phosphate there is a shift in shape that makes the fit with the actin much more effective (1); when Brownian motion brings the two into contact the myosin binds strongly to the filament (2).

It is the release of the phosphate ion from the binding site that provides the power stroke. Compare the shape of the myosin head in 1 and 2 (yellow) to that in 3 and 4 (red). Black lines have been added to 2 and 3 to indicate the change. When the phosphate ion is released, a small change in shape occurs at one point in the structure that is amplified into a much larger movement by another part of the structure acting as a lever arm. As a result, the central portion of the molecule switches from a bent to a straight form, which has the effect of yanking the actin filament forward by several nanometres, as indicated by the arrow in 3. The cycle is repeated when the bound ADP is replaced by ATP and the molecule detaches from the track (4).

Notice that in this case the protein detaches completely from the filament at one stage in every cycle. While it is detached, diffusion can help to push it along and the shapes of the molecules are such that one direction is favoured over the other. This increases the length of each step taken by a myosin motor protein, and hence its speed. The lever-arm section of the molecule is based on a rigid α-helix of variable length, which can swing through an angle of up to 70°. From a tiny change of only 0.5 nm, brought about by the addition or removal of a phosphate group, steps as large as 36 nm have been measured, from a combination of a 25 nm power stroke and an extra 11 nm contribution from diffusion.

The forces generated by a molecular motor are of the order of just a few piconewtons, pN (pico = 10^{-12}). This is more than enough for shifting vesicles and other small objects around cells, and different kinds of linkage have evolved to bind to different types of cargo. For larger-scale motion the cooperative action of many motors is required.

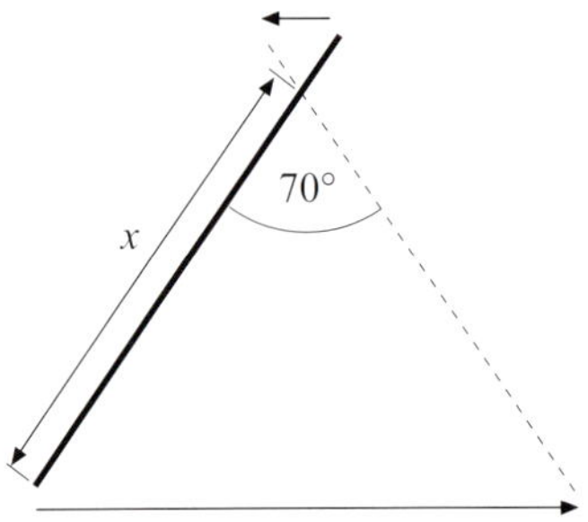

Figure 3.46 Myosin lever arm

SAQ 3.20 (LEARNING OUTCOME 3.5)

(a) A kinesin motor takes 8 nm steps, with each step powered by the conversion of one molecule of ATP to ADP. If the force generated by the motor is 2 pN, calculate the amount of useful work that can be done per step, and the efficiency of the motor.

(b) Estimate the number of myosin motors that would be needed to lift an apple.

(c) Figure 3.46 is a much simplified representation of the lever arm that operates within a myosin molecule, amplifying an initial movement of 0.5 nm to a power stroke of 25 nm. Find, by calculation or scale drawing, the length of the lever arm, x. How many turns of an α-helix does this represent? (The pitch of an α-helix is 0.54 nm.)

5.2 Muscle power

It should be obvious from your previous calculations that the large-scale motion achieved by muscles requires the cooperative action of many millions of molecular motors. Just to keep our hearts beating requires a constant power output of up to 5 W, while other muscles can deliver much higher power for more limited times: the quadriceps in the leg, for example, can produce over 100 W, but only for a few minutes. It is easy to calculate, but much harder to comprehend, the rate of ATP conversion that this represents.

The body contains different types of muscular tissue, but the molecule at the heart of all muscular motion is myosin. Several hundred myosin molecules bind together to form thick filaments, and arrays of filaments work together within muscle tissue to generate the power required for locomotion, digestion, circulation and to keep our hearts pumping. We've already seen how each individual myosin produces a power stroke and Figure 3.47 shows how, in muscle cells, the tails from many myosin molecules bind tightly together to form filaments, with the heads protruding from both sides.

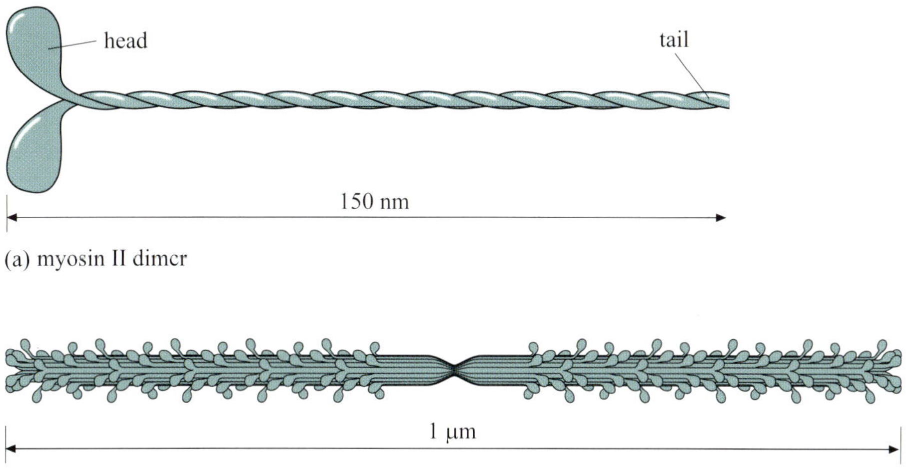

Figure 3.47 (a) A single myosin dimer and (b) a muscle myosin filament assembled from multiple dimers with their tails intertwined. Myosin II refers to a particular form of myosin

The heads of the myosin motors reach over and bind to actin filaments. Muscle contraction involves the relative motion of interleaving myosin and actin filaments, not unlike the interleaving fingers of the comb actuators you studied in Block 1.

Can we use the power of molecular motors outside the natural environment of the cell? It certainly seems that this might be possible, although a lot more research is needed before we fully understand how they work and how they might be adapted to other functions. It is known that at least some molecular motors can be persuaded to function outside the cell and you have seen some examples in the text.

You have already read about tiny propellers made to rotate by ATP synthase, and 'musclebots' that we can assume to be based on myosin. In both cases the protein motors were combined with non-biological materials to produce hybrid devices. Kinesin motors have also been isolated. In one demonstration they were stuck onto the flat surface of a microscope slide and a solution of microtubules was added. The kinesin molecules were able to push the microtubules around, like a cellular transporter working in reverse. Kinesin is a popular choice for nanotransport-system designers because it is readily available and can convert chemical energy from ATP to mechanical work with efficiencies of over 50%. SAQ 3.22 will give you an idea of the power that is available from such a system.

EXERCISE 3.12

Describe an example of a protein-based molecular motor used to generate motion outside the cell. Suggest an engineering application for molecular motors.

SAQ 3.21 (Learning outcome 3.5)

(a) What are the main similarities and the main differences between the structure and operation of the kinesin and myosin motors?

(b) The force needed to detach a myosin head from actin is approximately 10 pN. How does this compare to the tensile strength of the filament that you calculated in SAQ 3.19?

SAQ 3.22 (Learning outcome 3.5)

Kinesin motors can take about a hundred 8 nm steps a second, with a force of up to 7 pN. What power does this represent? How does this compare to the ATP-synthase motor?

6 MANUFACTURE

6.1 DNA: a load of old rubbish?

You probably know a little about DNA already. Since the first announcement of its structure by Watson and Crick in 1953, based on X-ray diffraction patterns from Rosalind Franklin, it has captured the imagination of scientists and artists alike. A few recent sculptural interpretations of the famous double helix are shown in Figure 3.48.

(a) (b) (c)

Figure 3.48 A selection of DNA sculptures: (a) fibre-optic illuminated DNA sculpture designed by Anne Arlidge and Jancis Smithells, for the new Royal Berkshire Hospital entrance lobby; (b) DNA sculpture that doubles up as a climbing frame, from the Lawrence Hall of Science, California, USA; (c) DNA from shopping carts, by Abigail Fallis, on display at 'Sculpture at Goodwood', Surrey, UK

You have probably also heard DNA described as the 'genetic blueprint', and indeed it is. Our DNA holds instructions for building all the proteins that our bodies need to function, in a remarkably compact and stable form. However, our DNA also contains a lot of rubbish: it is estimated that only two or three per cent of our DNA contains information for building proteins. Some of the rest is used to define and separate different sections, using standard repeating motifs. Other repeating sections are thought to arise from genetic parasites that use our DNA as a means of reproducing themselves, but serve no useful function in the cell. Much of the rest is simply a waste product of over three billion years of evolution. So, while marvelling at the elegant simplicity of the DNA double helix, it is also worth speculating on whether there may be ways in which this remarkable molecule might be used differently, or more efficiently.

In this section, I will start by showing you how DNA stores information and how that information can be retrieved and duplicated. I'll then consider how proteins are made using information from DNA, and how this process is itself mediated by proteins; you'll get a closer look at a few of the molecular machines at work here. Finally, I'll briefly introduce some of the possible future directions of DNA technology.

6.2 Information storage

Each strand of the DNA helix is a polymer constructed from a limited set of basic building blocks known as nucleotides: molecules made up of a phosphate group, a sugar, and a base. When the nucleotides link together, the sugar and phosphate groups form the backbone of the chain with the bases protruding from the sides, as described in Section 14 **Nucleic acids** in the *Principles* wing. For the current discussion the main points to remember are these:

More information about the structure of DNA (and RNA) bases and how they link together to form a single chain and a double helix is given in Section 14 Nucleic acids in the *Principles* wing.

- DNA contains four different bases: adenine (A), guanine (G), cytosine (C) and thymine (T).

- Along each DNA chain the nucleotides are linked together by strong covalent bonds to create a sugar–phosphate backbone.
- The bases protrude from the sides and can link together via hydrogen bonds; the combinations of A with T, and G with C, are strongly favoured.
- The DNA molecule is a double helix, held together by interactions between bases on adjacent chains.

A simplified picture of the DNA structure, showing how the chains are linked by hydrogen bonding between bases, is shown in Figure 3.49. Note that G and C are linked by three hydrogen bonds, while A and T are linked by only two.

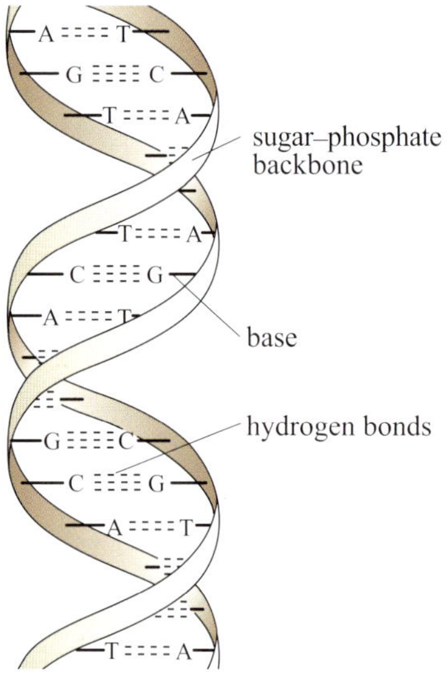

Figure 3.49 The double-helical structure of DNA, showing hydrogen bonding between the bases on the two strands

Consider how information is stored digitally in your computer. The computer uses a binary code where all the information is stored as a combination of just two digits, 1 and 0, known as 'bits'. A set of 8 bits, known as a byte, is used to represent a single character in the computer's memory. The hard drive of a computer uses magnetic means for storing data and the very best coverage achieved at the time of writing is about 100 Gbits in^{-2} (as the Americans would say) – which, translated into more sensible units, I make about 6.5×10^3 nm^2 bit^{-1}. This makes the side of each bit less than 100 nm in length.

For each position on a DNA strand there are four possible values: the four bases A, G, C and T. The equivalent of the byte is a sequence of 3 nucleotides called a *codon*, since the sequences provide a code for describing proteins. There are 64 possible codons: AGC, ACT, GCT, etc. Most of these sequences are assigned to one of the 20 naturally occurring amino acids, while a few others are used to give directions like 'start' or 'stop'. For some amino acids there are several possible codons; for instance, any of the four 3-letter combinations starting with GG translate to the amino acid glycine. Taking account of the atoms needed to position each base correctly, the density of information storage is estimated to be about one bit per 15 atoms: not bad!

There are higher levels of organization as well. A section of DNA that encodes all the information required to make a particular protein is a gene; genes are often considered to be the basic units of heredity. Genes are grouped together, along with other DNA sequences, into bigger units called chromosomes, large enough to be seen under an optical microscope. Humans have 23 pairs of chromosomes, one set from each parent. The sum total of all the genetic information of an organism is called the genome: the human genome contains approximately 3×10^9 base pairs.

SAQ 3.23 (Learning outcome 3.8)

(a) Explain what is meant by a codon. Why are three letters needed for a codon – could a suitable code be devised using only two?

(b) Show that a data storage density of 100 Gbits in^{-2} corresponds to a bit size of approximately 6.5×10^3 nm^2. (1 in = 25.4 mm.)

(c) Assume that the atoms involved in storing a bit of data in DNA can be modelled by a grid of 4×4 carbon atoms, where the distance between the centres of neighbouring carbon atoms is 0.2 nm. Calculate the area occupied by a bit and compare it to the value above.

6.3 Reading the message

In order to use the information stored in DNA to make the proteins necessary for survival, and to grow and reproduce, cells need methods for copying and interpreting the code. The process of copying DNA, essential for reproduction, is called *replication*.

To make proteins an intermediary is used to transfer the information. This is RNA: a second nucleic acid with a chemical structure very similar to DNA, described in more detail in the section on nucleic acids in the *Principles* wing. The different name is derived from a change in the sugar component in the backbone of the chain; ribose in RNA replaces deoxyribose in DNA. The other major difference is that the base thymine (T) in DNA is replaced by uracil (U) in RNA. The process of copying the information from DNA to RNA is called *transcription*.

DNA replication and the transcription of DNA into RNA take place within the nucleus of a cell. If you look back at the diagram of an animal cell in Figure 3.1(a) you will see that the nucleus is surrounded by a convoluted system of membranes labelled endoplasmic reticulum (ER), which may be rough or smooth. It is here that proteins are synthesized, in a process known as *translation*, and folded up into the correct conformation. Translation is an appropriate term; it is at this stage that the chemical language of the four nucleotides in nucleic acids is read and translated into the very different chemical language of the twenty naturally occurring amino acids in polypeptides.

Translation is made possible by *ribosomes*: complex molecular machines built in the nucleus from protein and RNA molecules, which assemble protein chains according to instructions read from RNA. Ribosomes are visible as small bumps on the surface of the rough ER, or may float freely within the cell. The smooth ER shown in Figure 3.1 has other roles in synthesis, not covered here.

To use DNA effectively cells have evolved a toolbox of enzymes that interact with the chain in different ways. The main types, and their functions, are listed in Table 3.5. Varieties of most of these enzymes are available commercially and are equally effective outside the cell. Using enzymes such as these to copy and manipulate DNA provides the basis for genetic analysis, which has many applications in diagnostics and forensic science. We can also use enzymes to modify DNA, providing the tools for creating custom proteins and engineered organisms through genetic engineering as well as other constructions unlike any found in the cell. Of course, having the tools is one thing – knowing what to make is quite another!

It would be useful to remember the key functions from this list, but there is no need to memorize the names of the enzymes.

Table 3.5 Enzymes used to copy and manipulate DNA, both inside and outside cells

Enzyme family	*Function*
Helicases	Temporarily separate the two strands of a DNA double helix so that replication can take place.
Polymerases	Use information from one strand of nucleic acid to polymerize a new strand.
Nucleases	Cut nucleic acid strands at specific points along the chain. Base-excision nucleases remove individual bases and are used to correct errors.
Ligases	Connect nucleic acid strands together.
Regulatory proteins	Regulate the use of information, e.g. by blocking part of a strand or guiding a polymerase to a particular point.
Topoisomerases	Deal with topological problems, e.g. by unwinding the helix by 'nicking' the chain, allowing it to swivel and then resealing it.
Recombinases	Extract a portion from one strand of DNA and swap it with a similar portion in a different strand.
Spliceosomes	Remove a portion from within a strand of RNA and splice the two ends together.
Nucleosomes	Along with other proteins, package nucleic acids for storage.

6.3.1 Copying DNA

DNA replication – the copying of the DNA chain – starts with the separation of part of the double helix into two separate strands, each of which can be used to generate the matching half of a new chain. Preferential base pairing is the key and the process is catalysed by the enzyme DNA polymerase.

The process of replication is illustrated in Figure 3.50. The two strands of the double helix unwind, starting at one end, to expose the sequence of bases on each strand. Each of the individual strands now provides a template for a new strand, constructed so that the base-pairing rules are followed: satisfy yourself that this is the case in Figure 3.50(c). As nucleotides with complementary bases are added along each chain, covalent bonds form between the sugar and phosphate groups in adjacent units to form the backbones of the new chains.

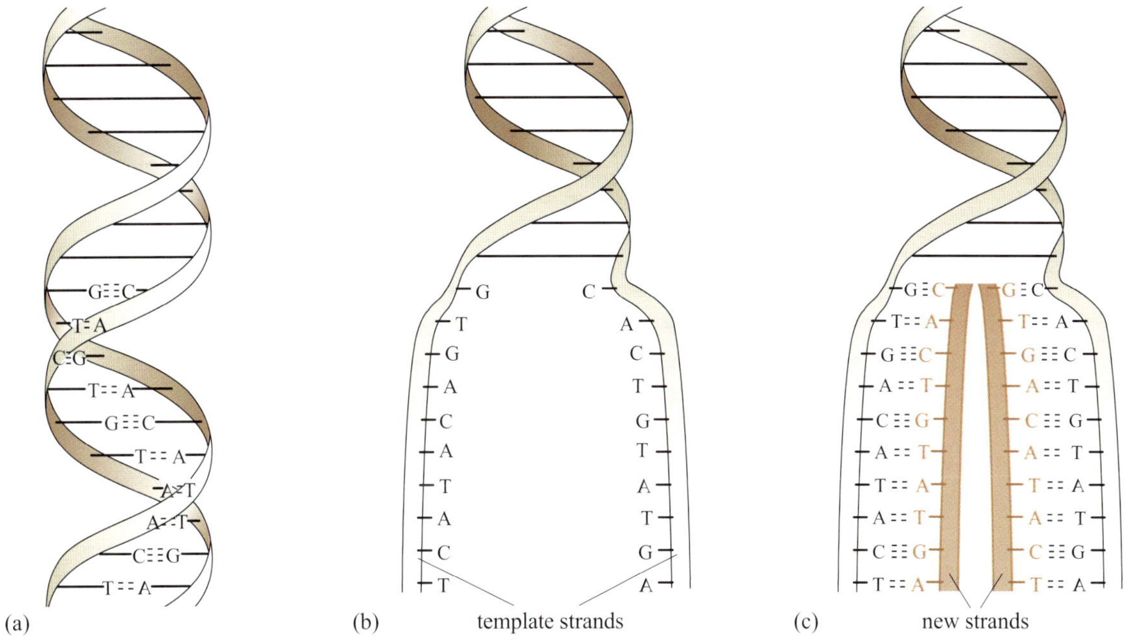

Figure 3.50 The process of DNA replication: (a) a portion of a DNA double helix; (b) part of the double helix unwound; (c) part of each chain has been replicated, and two new strands have been produced. The two paired chains have not yet wound into double helices. The process continues until the entire parent molecule has been replicated

The result is the production of two identical strands of DNA, each of which contains half of the 'parent' chain. At least, that's the plan. In reality mistakes do occur. Some of the proper pairings are only slightly more favourable than some of the improper ones: for instance, a T may be added where a C should have been. DNA polymerase incorporates a 'proof-reading' step to the process, which checks each pairing as it is produced. It can do this because the molecule has two active sites. One site matches each base with its partner on the template strand and connects it to the new growing strand. The other site then pulls on the new base, and if it is held less tightly than expected, it cleaves it off. T paired with G can form only two hydrogen bonds, instead of the three expected from a (correct) C–G pairing, so it would be removed. This extra quality-control procedure increases the accuracy considerably, to as little as one mistake per billion bases; yet this is still enough to allow gradual changes to take place that can lead to evolution through natural selection. All DNA polymerases are the right shape to wrap around DNA; the most effective also incorporate a 'clamp' component, which allows the enzyme to move along the DNA as it copies, but prevents it from detaching completely.

The transfer of this natural copying process to the test tube was a major breakthrough, allowing tiny amounts of DNA to be rapidly multiplied to produce enough sample for analysis. Development of the ►***polymerase chain reaction***◄ (PCR), by which this is achieved, won the Nobel Prize for Chemistry for its creator, Kary Mullis, in 1993.

▶ The polymerase chain reaction ◀

The polymerase chain reaction (PCR) is sometimes referred to as 'molecular photocopying' because it provides a fast and inexpensive method of reproducing copies of DNA. The PCR makes use of a particularly stable form of DNA polymerase isolated from bacteria that live in hot springs and have therefore evolved enzymes that can survive elevated temperatures. The basic process is illustrated in Figure 3.51.

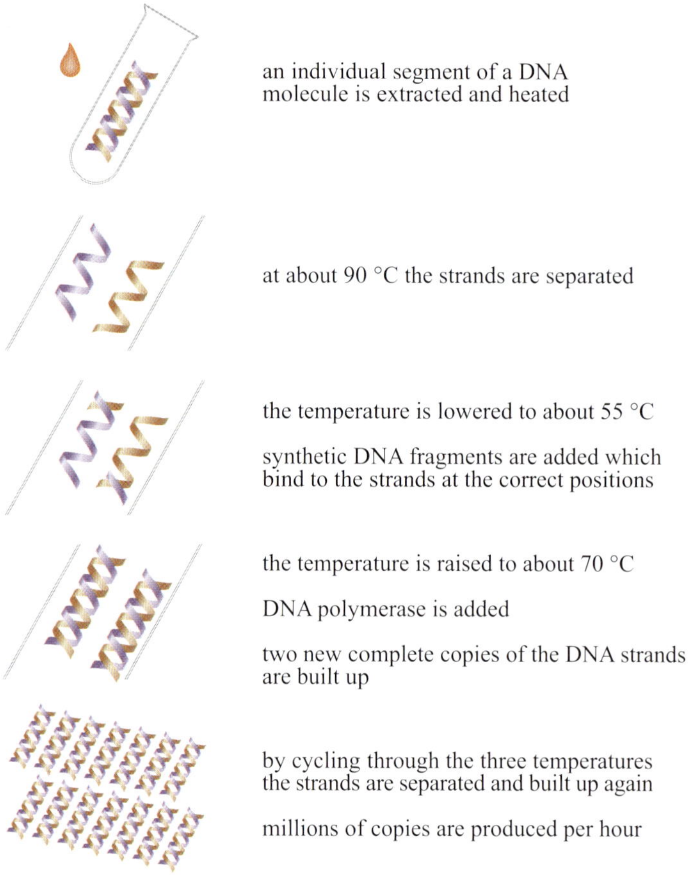

Figure 3.51 The polymerase chain reaction

The PCR provides a good example of how natural processes can be adapted for uses outside the cell. Replication is speeded up by cycling through different temperatures to repeatedly separate strands of DNA and then build them up again, doubling the quantity on every cycle. Large quantities of DNA are produced in just a few hours, providing the raw material for techniques such as DNA fingerprinting, detection of bacteria or viruses, and diagnosis of genetic disorders.

6.3.2 Analysing DNA

Cells replicate DNA in order to grow and multiply, but outside the cell our motives may be quite different and we like to know what we have got. Having made multiple copies of a piece of DNA, how do we then find out what is in it? One common approach to reading the DNA code is the *chain termination method*. In this procedure the chain is first chopped up into small fragments. The four nucleotides A, G, T and C are added to a solution containing many single strands of the DNA under test, together with primers to initiate copying. A small proportion of the nucleotides have been modified so that when they are added to a chain, they block the end and prevent further extension. Traditionally four different experiments would be carried out, one for each of the four bases. In modern versions of the technique the four different 'blockers' are labelled with coloured markers, so that they can be easily differentiated. All four can then be tested simultaneously, and the results read by a laser scanner.

Let's see how this works by considering a short portion of a chain of single-stranded DNA just twelve bases long. Suppose the sequence of my original segment is ATACCGTACTGA, so that a complete copy would result in a chain with the sequence TATGGCATGACT. Suppose also that the primer attaches itself to the first two groups, so that every copied chain begins with TA. I will label my modified nucleotides so that they appear with these colours: A (orange), G (blue), T (purple), C (yellow). Remember that these coloured groups stop the chain from growing any further. Considering each base in turn, the chains that could result from this mixture are shown in Figure 3.52(a); they are arranged in order of length in Figure 3.52(b). This would be read by the laser scanner as represented in Figure 3.52(c) and, once the two known bases at the start are included, is easily translated to reveal the original sequence of the DNA. As a result of technology like this, the complete sequence of the 3×10^9 base pairs that make up the human genome has been determined and was published in 2001.

	TATGGCA	TATGGCATGA	
	TATG	TATGG	TATGGCATG
	TAT	TATGGCAT	TATGGCATGACT
(a)	TATGGC	TATGGCATGAC	

TAT
TATG
TATGG
TATGGC
TATGGCA
TATGGCAT
TATGGCATG
TATGGCATGA
TATGGCATGAC
(b) TATGGCATGACT

(c) 

Figure 3.52 Chain termination method applied to DNA sequence ATACCGTACTGA: (a) copied chains produced, showing 'blockers' labelled with coloured markers; (b) copied chains arranged in order of increasing length; (c) a representation of the markers as read by the laser scanner

EXERCISE 3.13

Deduce the colour sequence that would emerge from a chain termination analysis of the DNA sequence GATACCTGAACT, making the same assumptions as above.

The method most commonly used to separate a mixture of DNA fragments according to their length is *electrophoresis*. In this technique an electric current is passed through a medium – typically some sort of gel – containing the mixture, and each fragment travels through the medium at a different rate, depending on its electrical charge and size. Separation is based on these differences. Electrophoresis is not just useful for DNA analysis but can be used to separate other large molecules, such as proteins, from a mixture.

Having read and translated parts of the code, it is often useful to look for a particular sequence of bases in a piece of DNA. For instance, it may be possible to link errors in the DNA sequence of a particular gene to a certain genetic disorder. For this purpose we can use tiny grids of capture probes – microarrays – in much the way as discussed previously for sensors based on the detection of specific proteins.

DNA microarrays consist of short stretches of DNA attached at one end to a solid surface such as glass or silicon: these are the capture probes, and stretches of DNA from the sample under test will bind to them if they have the correct complementary sequence. An example is shown in Figure 3.53. In this example the correct base sequence is TATAGCCG, which will bind to the capture probe with the complementary sequence ATATCGGC. In the defective gene the sequence has mutated to give TATACCCG, which will bind to a different capture probe.

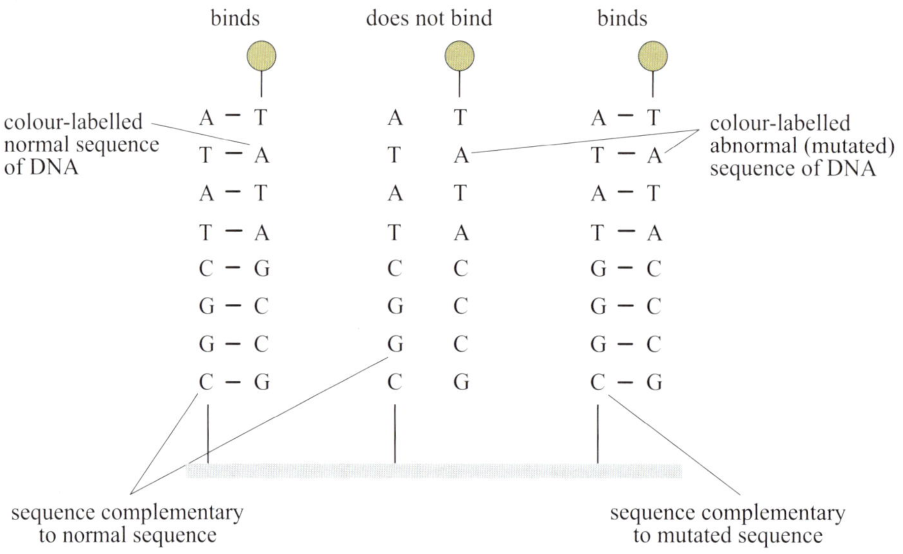

Figure 3.53 Example of DNA capture probes used to test for normal and mutated DNA sequences

The DNA sample under test can be amplified using the PCR, and labelled with a fluorescent marker. By making microarrays with many different capture probes in known positions, it is possible to test for many different sequences at the same time and read off the results from the pattern made by the fluorescent markers.

Procedures such as those above have been made much more accessible by the development of tiny analysis systems known as labs-on-a-chip, or micro-total-analysis systems. As the name suggests, these are miniaturized versions of laboratory procedures that can combine several stages of sample preparation, analysis and detection in one small area, with sample volumes down to a few nanolitres. Special design features and fabrication methods have been developed to work with liquids at this small scale: many of these are derived from the microfabrication methods of the electronics industry that you have studied in previous blocks, adapted to suit the particular requirements for interacting with biological systems.

Lab-on-a-chip technology has many advantages over traditional methods, providing potentially low-cost, high-throughput, portable and efficient devices with minimal environmental impact because of their low energy and chemical requirements. Such devices provide the practical means for exploiting many of the lessons we can learn from nature and putting them to good use. You can read more about **lab-on-a-chip** and **electrophoresis** in the *Techniques* wing.

More details about the technology behind lab-on-a-chip miniature analysis systems are given in Section 8 Microfluidics and lab-on-a-chip in the *Techniques* wing.

Section 9 Electrophoresis in the *Techniques* wing describes how large molecules can be separated from a mixture according to their size.

SAQ 3.24 (Learning outcomes 3.8 and 3.10)

Suggest a possible design for a sensor incorporating capture probes to detect a particular mutated gene sequence of GCTTACGGA.

6.3.3 *Building new proteins*

Returning to the cell, the next stage in the journey from DNA to protein is the synthesis of RNA using DNA as a template. This process is illustrated in Figure 3.54: the similarity between this and Figure 3.50 is very obvious. You'll notice some slight differences in the base pairing because of the replacement of T with U on RNA, and another key feature is that only one chain of RNA is synthesized: in fact, RNA usually exists as a single-stranded chain.

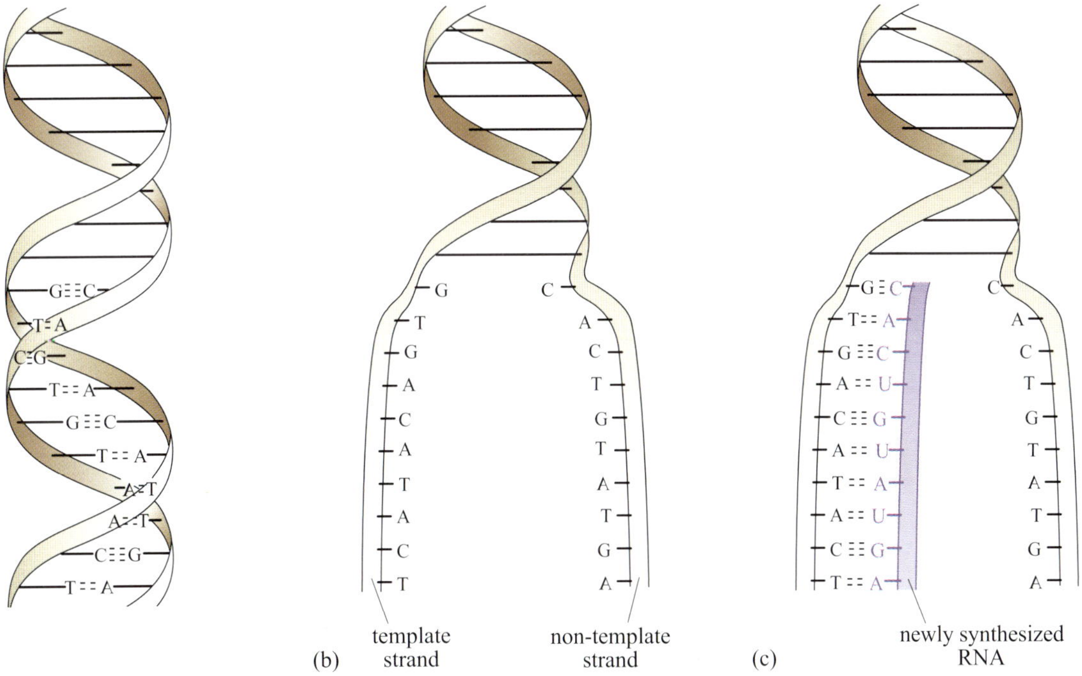

(a) (b) (c)

Figure 3.54 Transcription: the synthesis of RNA on a DNA template. (a) The DNA double helix; (b) the hydrogen bonds between the base pairs have been broken and the two strands have separated – note that only one of the strands is used as the template for RNA synthesis; (c) a short length of RNA (shown in purple) has been synthesized. In reality the RNA molecule would be much longer than this, but much shorter than the whole DNA chain

Another difference, not obvious from the figure, is that only a small section of the DNA is copied at a time. The section to be copied corresponds to a gene, and its position is defined by 'start' and 'stop' codes on the DNA chain. Errors in transcription are less serious than those in replication (a defective strand of RNA may not make its protein correctly, but since the original DNA is still available a replacement can always be made) so the proof-reading at this stage is less rigorous.

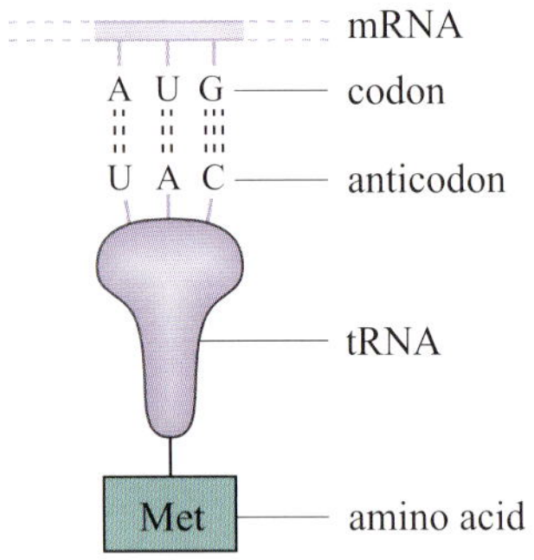

Figure 3.55 Stylized representation of tRNA molecule providing translation from the mRNA codon AUG to the amino acid methionine (Met)

Three different types of RNA are made, for three different purposes:

- The *blueprint* for building a protein chain is carried by messenger RNA (mRNA).
- *Translation* from the language of nucleic acids to the language of proteins is achieved using transfer RNA (tRNA).
- *Assembly* of new proteins requires ribosomal RNA (rRNA).

Figure 3.55 shows a stylized representation of a molecular translator, a tRNA molecule. The middle section is a folded-up strand of RNA. At one end of the chain is a dangling end, which provides a binding site for a specific amino acid. On the other side of the molecule (at the top of the figure) there are three unpaired RNA bases: these form the *anticodon*, which has the base sequence complementary to a codon on mRNA.

In Figure 3.55, the message carried by the mRNA codon is AUG, which codes for the amino acid methionine (Met). The tRNA has the complementary base sequence UAC at its anticodon, and carries a methionine unit bound to its loose end.

We now have the blueprint for a protein, and the ingredients for making it. The molecular machine that actually makes the protein is the ribosome: a true molecular assembler. If we wish to make self-replicating protein-based devices, then we need a machine to assemble them. We need a ribosome.

An animated version of Figure 3.56 has been provided on the course DVD.

The ribosome reads instructions from the mRNA blueprint, translates them using tRNA, and uses them to construct proteins, one amino acid unit at a time. Electron microscopy has proved particularly useful in working out how this machine works, and detailed structures are available in the Protein Data Bank. Figure 3.56 shows a highly schematic representation of a ribosome in action.

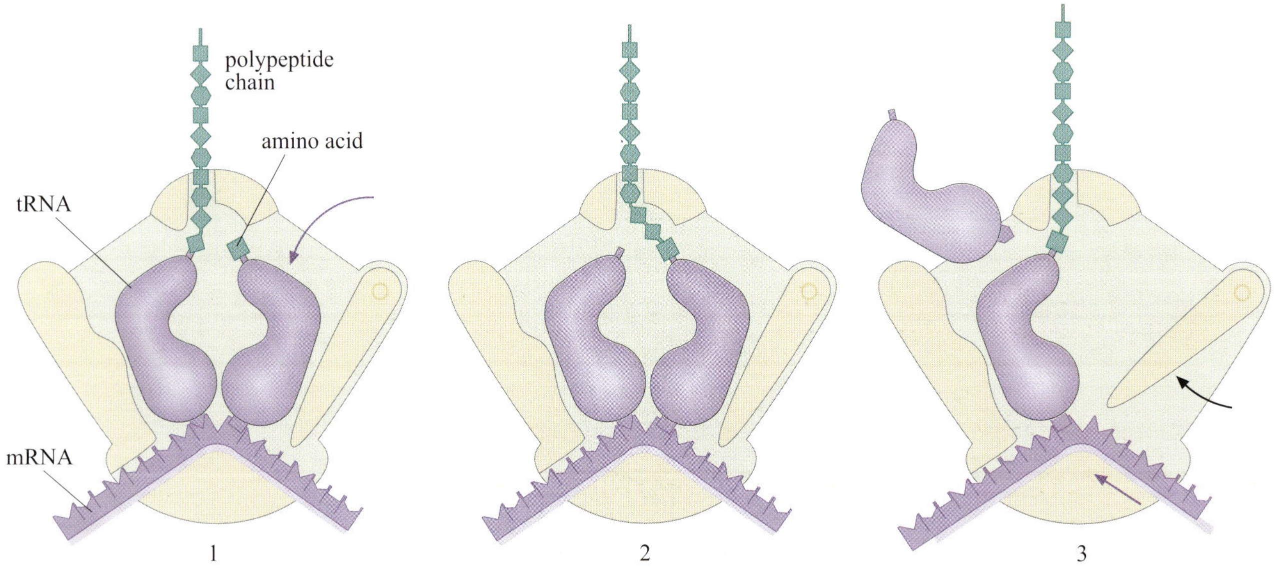

Figure 3.56 The ribosome steps down the mRNA strand one codon at a time

The ribosome clamps itself around the mRNA chain that carries the instructions for protein synthesis, and steps down the mRNA chain one codon (3 units) at a time. A new tRNA molecule enters at the right with the correct match to bind to the chain (1). The corresponding amino acid is attached to its other end. The growing protein chain is transferred to the newly arrived amino acid held by the tRNA (2). Finally the tRNA is pushed one step forward – ejecting the previous, now empty, tRNA (3). In this way the protein chain is progressively built up, one unit at a time. Needless to say, many other enzymes are involved to help the process along.

Translation is a much less accurate process than replication or transcription. In bacteria it is estimated that about one in four proteins of typical length will contain an error, although many of these won't prevent the protein from functioning properly. It has been suggested that the use of RNA as an intermediary between DNA and proteins is an evolutionary legacy and that RNA was developed first, with DNA appearing only later as a means of archiving complex information. Some simple organisms rely entirely on RNA to carry their genetic blueprint.

The final steps in protein synthesis involve folding the chain up into the correct shape and delivering it to where it is needed in the cell. A single molecule of a globular protein may have a preferred low-energy conformation when alone, but if other molecules are present then alternative energetically favourable structures may well be achieved through chains coagulating. Remember that lipids form self-assembled structures, so that their hydrophobic regions are clumped together and sheltered from the surrounding water. Protein folding is also driven largely by the need to shelter hydrophobic groups from water; clumping could achieve the same result, but would leave us with a useless product. A class of proteins known as *chaperones* have the job of taking newly synthesized polypeptide chains and keeping them separate from each other in a suitable environment for them to fold into useful protein structures.

So DNA makes RNA, which makes proteins that carry out just about every function required by cells. But there are other possible uses of DNA that are not exploited by nature. I'll finish this section by looking at a few of these.

SAQ 3.25 (Learning outcome 3.8)

(a) Describe how you would expect the conformation of a chain of mRNA to differ from that of tRNA.

(b) Assuming that each tRNA molecule can bind specifically to only one kind of amino acid, what is the minimum number of different tRNA molecules required by a cell?

(c) What base-pairing rules must apply to the interaction between the codon and the anticodon?

(d) One codon representing the amino acid glycine (Gly) is GGA. Draw a sketch similar to Figure 3.55 with Met replaced by Gly.

SAQ 3.26 (Learning outcome 3.8)

Suppose you were to design a system to synthesize proteins directly from DNA. What changes would you need to make to the ribosome and its associated molecules? Suggest possible advantages and disadvantages of such a system.

6.4 Other applications for DNA

DNA has potential as a self-assembling structural material, and we have a well-equipped toolbox of enzymes available to manipulate it. The double helix is stiff over lengths of up to about 10 nm. Chains can be made with selectively sticky ends, so that different components can be fitted together by designing them with complementary sequences of unpaired bases. In this way both static structures and functioning DNA-based machines can be constructed.

Figure 3.57 shows examples of two simple structures built from DNA. The two-dimensional network on the left is constructed from a four-armed module – one is shown in a lighter shade in the centre – with sticky ends designed to link the units together. The three-dimensional cube on the right (with one face picked out in a lighter shade) is created from eight interlocking three-armed modules. Further components can be added by the incorporation of dangling ends into the structures. It has been suggested that constructions such as these could provide the basis for building circuits for nanoelectronics, or for data storage.

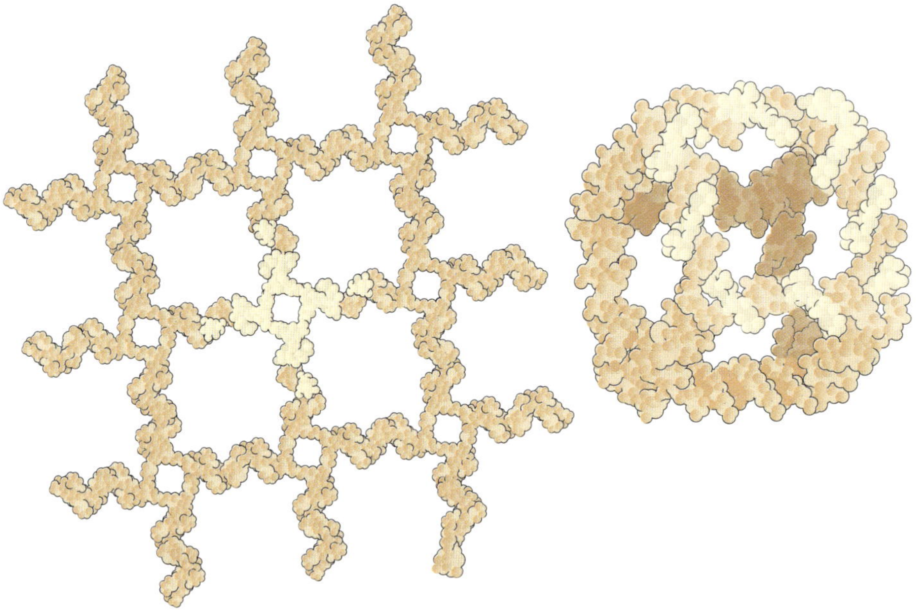

Figure 3.57 DNA networks created by linking modular units together through self-assembly

A moving device – a pair of molecular tweezers – is illustrated in Figure 3.58. The tweezers are made from three short strands of DNA, partially twisted together as shown in the upper picture. There is a short unpaired segment in the centre that acts as a flexible hinge, and an unpaired stretch of chain at each end. The fourth strand is designed to bind to these two unpaired stretches of chain, pulling them together and closing the tweezers. Different molecules can be attached at the points marked with stars and thus brought together when the tweezers close.

More sophisticated devices have also been constructed, including a pair of DNA legs that can walk across a surface coated with DNA strands using the successive binding and detaching of complementary base pairs.

The main role of DNA in the cell is information storage. It has been suggested as a candidate for very long-term storage of data because, although the slow speed of reading the data makes it impractical for everyday use, it is a technology that we can

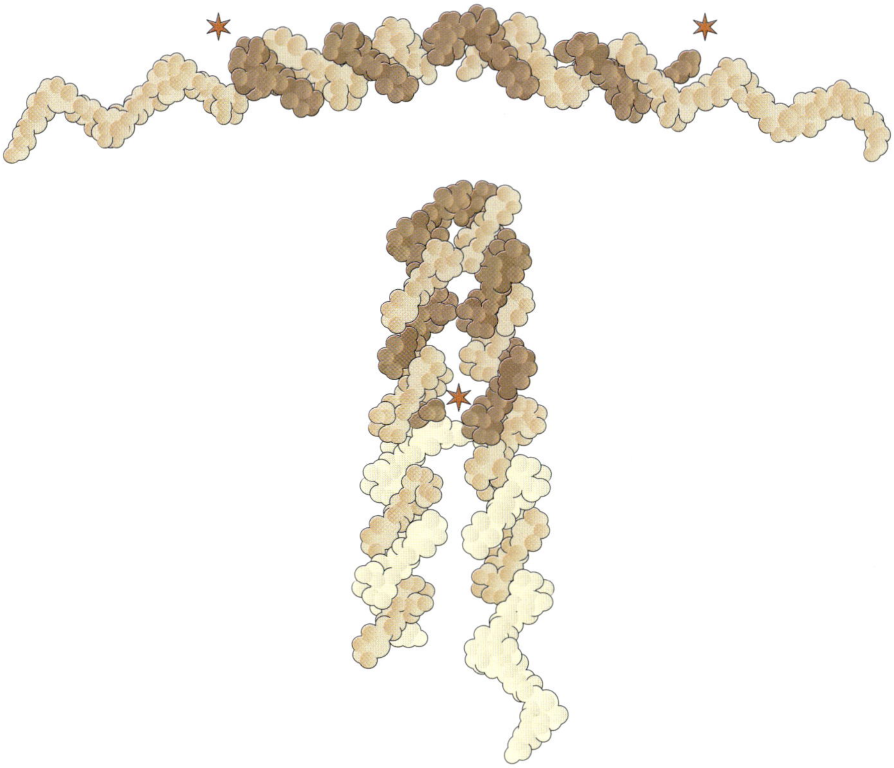

Figure 3.58 DNA tweezers

reasonably expect to be in use even after hundreds or thousands of years – unlike current data-storage methods, which seem to become obsolete with alarming speed! However, cells make very little use of the potential of DNA for computation. Various mathematical problems have been successfully solved using DNA, using the pairing of complementary strands as the basis of the method and testing many possible solutions to a problem simultaneously, simply by mixing them together in a test tube. One simple example is given in ►***Computing with DNA***◄.

The idea of using DNA in this way to solve problems is intriguing but not, for now at least, terribly practical; apart from the inevitable errors in copying, a completely new set of DNA strands must be created for each individual problem, which makes it both time-consuming and expensive. Perhaps a better role for DNA in the development of computing is as the basis for nanoscale conductive wires.

Some researchers claim that DNA can itself be used as a conducting wire, but this is very difficult to achieve. However, the negatively charged phosphate groups along the chain, which are normally associated with sodium ions, can act as a substrate for the attachment of silver ions. Tiny wires have been made by connecting a length of DNA between two gold electrodes with capture strands attached to them and exposing it to a solution of silver nitrate. The silver ions associate with the chain and then act as nuclei for the growth of larger silver grains that eventually merge to form a wire about 100 nm wide. Finally, the wires are coated with gold. In a process described as 'molecular lithography' some sections of the chain can be blocked using proteins, which act as a resist and provide sites on the wire where other components can be added. This opens up possibilities for creating electronic circuits on much smaller scales than is possible using conventional lithography.

▶ Computing with DNA ◀

In this problem DNA is being used to find all the 3-digit binary sequences x_1, x_2, x_3 that satisfy these two conditions:

Condition 1: either $x_1 = 1$, or $x_2 = 1$, or $x_3 = 0$

Condition 2: either $x_1 = 0$, or $x_2 = 1$, or $x_3 = 0$.

There are eight possible sequences of three binary numbers, and each is allocated to a different string of five DNA bases, as shown in Figure 3.59. Capture strands with these sequences are attached to a surface, which is then exposed to different solutions containing complementary base sequences.

binary string x_1	x_2	x_3	DNA string	surface
0	0	0	ATGCC	1
0	0	1	TGCGG	2
0	1	0	AAGCG	3
0	1	1	CCTAT	4
1	0	0	TAGAC	5
1	0	1	GGATT	6
1	1	0	CTTCG	7
1	1	1	GTAAT	8

Figure 3.59 Using DNA for problem solving

First, all the complementary strands that satisfy condition 1 are added to the solution; they will stick to the strands numbered 1 and 3–8. The remaining unpaired strand (number 2) is destroyed by enzymes. The complementary strands are now removed and washed off the surface and a new set of test strands is added, which represent all the combinations that satisfy condition 2. This time strand 6 is left unpaired and destroyed. The six strands that remain give all the possible combinations that satisfy both conditions. They can be removed, amplified using the PCR, and identified against an array of test strands of known sequence. In total, this requires three steps for each condition to be tested plus one to analyse the result: for k clauses this is $(3k + 1)$ steps. The number of steps required is the same even if the number of variables (values of x) is increased.

This may seem rather a long-winded approach for such a trivial problem, but as the number of variables and the number of clauses are increased this method takes considerably fewer steps than a computer would need to work out the same result. For example, a problem with 30 clauses and 50 variables could be solved in approximately 1.6 million steps by ordinary computing methods but in only 91 steps by a DNA computer like the one described.

SAQ 3.27 (Learning outcome 3.4)

Suggest a strategy for making a circular loop of DNA double helix with circumference equal to the length of thirty-six DNA bases, from six individual strands. There is no need to specify all the base sequences you would use, but explain how you would apply both positive and negative design considerations when deciding which sequences to use.

EXERCISE 3.14

Suggest an engineering application for a construction built from DNA.

7 REVIEW

7.1 Molecular machinery

Look back again at the two representations of a cell in Figure 3.1. I hope that the biologist's view of the cell will now seem a little more familiar, and that you will be able to see the links between this and the engineer's view alongside it. We haven't covered all the functions in the cell, but you have seen examples of proteins acting as structural materials, catalysts, sensors, valves, pumps and motors, and seen how they are made from a blueprint carried by DNA. You've also learned a bit about lipids and nucleic acids and their roles in structural integrity and information storage.

There are many other structures and devices within cells that we could have covered. Those included here provide a good representative selection and serve to illustrate the basic principles of self-assembly, of binding sites and ligands, of the importance of shape and shape changes for the operation of molecular machinery, and of the types of energy transfer that are appropriate at this scale.

I have also aimed to give you a taster of some of the ways that the principles learned from biology, as well as some of the molecular machinery from cells, are being used in the development of new nanodevices. The *Techniques* wing and the journal articles from the course DVD will give you a more comprehensive view of a few selected areas. The following questions will help you to check on what you have learned.

SAQ 3.28 (Learning outcome 3.9)

Where in the cell would you expect to find each of the following? Describe their main function and classify them under one or more of the following general areas of operation: structures, communication systems, power production, transportation and manufacture.

(a) an electron transport chain

(b) phospholipids

(c) microfilaments

(d) ribosomes

(e) kinesin motors

(f) proton gradients

(g) sodium and potassium ion channels

(h) RNA

(i) ATP synthase

(j) axons.

SAQ 3.29 (Learning outcomes 3.3, 3.5, 3.6 and 3.8)

Give examples from the text of:

(a) structural hierarchy in a protein

(b) flow of electric current within a cell

(c) the conversion of
    (i) chemical energy to mechanical work
    (ii) electrochemical to chemical energy
    (iii) light energy to electrochemical energy
(d) rotational motion powered by a molecular motor
(e) linear movement powered by a molecular motor
(f) three ways in which enzymes can modify a DNA chain
(g) a use of DNA other than for storage of information
(h) a protein acting as a catalyst.

7.2 A note of caution

It is easy to get caught up in the wonder of biological devices and the potential they offer. In this block I have inevitably concentrated on a small number of individual elements, the structure and function of which are relatively well understood. But biology is a complicated business and I have barely scratched the surface of all the complex and interrelated processes that go on within cells. Many naturally occurring toxins and pathogens operate by interfering with the workings of the cell, for instance by physically blocking channels, disabling enzymes, or allowing natural processes to multiply out of control. We employ similar strategies ourselves for fighting disease and killing organisms we regard as pests or weeds. It is essential that any nanodevices that we modify or even invent from scratch do not have unforeseen consequences for ourselves or the wider environment.

The 'grey-goo' scenario, which speculates on the accidental (or perhaps deliberately malicious) creation of a plague of self-replicating robots, capable of spreading across the planet consuming its resources and threatening life as we know it, may currently belong in the realms of fantasy. However, it is a clear indication of the public concern, even fear, that surrounds much of this area. Genetic engineering is already a reality and has led to widespread ethical and political debate about where we should draw the boundaries of acceptable use.

There is a constant need not only for stringent control of new developments and rigorous testing of new products, but also for accurate public information and informed debate. The Royal Society report 'Nanoscience and nanotechnologies: opportunities and uncertainties' that you saw at the beginning of this course is a good starting point, and more will inevitably follow. There are interesting times ahead.

LEARNING OUTCOMES

After you have completed Block 3 main text you should be able to demonstrate that you have achieved the following learning outcomes.

You should be able to:

3.1 apply appropriate physical models to describe the behaviour of biological and biologically inspired systems at small scales

3.2 describe and give examples of how self-assembly enables construction 'from the bottom up' in natural materials

3.3 describe how a combination of strong and weak bonding within biopolymers and lipids is used to build hierarchical structures with common structural elements and finely tuned properties, including calculations where appropriate

3.4 explain how both positive and negative design principles must be applied to the design of molecular devices and comment on the challenges involved in attempting such design

3.5 explain the importance of molecular shape to the functioning of proteins, including description of how small changes in molecular conformation can be translated into larger-scale movement and mechanical work by molecular motors

3.6 describe the ways in which electrical signals are transmitted within cells and contrast these with conventional electric current

3.7 describe how energy is stored within cells in chemical or electrochemical form: give examples and carry out calculations to demonstrate how such energy is stored and utilized

3.8 account for the role of DNA as a means of information storage; explain how DNA sequences can be detected, read, interpreted, and used to build proteins; give examples of other potential roles for DNA outside the cell

3.9 identify the parallels between the molecular machines that operate within cells and more familiar large-scale engineering devices; describe the cell as an engineering system encompassing structures, communication systems, power production, transportation and manufacture

3.10 give examples of the application of biological principles to the design of practical engineering devices, including calculations where appropriate.

ANSWERS TO EXERCISES

EXERCISE 3.1

The one I had in mind was fermentation of yeast to produce alcohol, but you might also have come up with a similar use of yeast in baking to produce the carbon dioxide that makes bread 'rise'. Other possibilities could include the whole of agriculture, forestry to produce timber, some drug manufacturing routes, or many others. And where do you think coal, oil, paper, wool, cotton, natural rubber and leather come from?

EXERCISE 3.2

The conditions I thought of were: temperature of ~37 °C, atmospheric pressure, a water-based environment, constant pH, a constant supply of oxygen and a regular supply of essential nutrients.

EXERCISE 3.3

We know that:

$$\frac{3}{2}kT = \frac{1}{2}mv^2$$

Thus:

$$v = \left(\frac{3kT}{m}\right)^{\frac{1}{2}}$$

$$= \left(\frac{3 \times 1.38 \times 10^{-23}\,\mathrm{J\,K^{-1}} \times 293\,\mathrm{K}}{32 \times 1.66 \times 10^{-27}\,\mathrm{kg}}\right)^{\frac{1}{2}}$$

$$= 478\ \mathrm{m\,s^{-1}}$$

To get a feeling for what this means, compare it to the speed of sound in air, which is 330 m s^{-1}.

EXERCISE 3.4

(a) The sequence of amino acids along the chain.

(b) There are 20 naturally occurring amino acids.

(c) At each position along the chain there are 20 possibilities. So the number of possible sequences is 20^6 = 64 000 000.

EXERCISE 3.5

The example I thought of was tissue engineering, where biological materials are being developed to provide substitutes for living tissue – for instance, to replace skin or to repair damaged blood vessels. An ultimate aim is to find ways to prompt the body to grow its own replacement parts.

EXERCISE 3.6

Hydrophobic side groups are more likely to be in contact with the lipid bilayer, i.e. on the outside surface of the α-helix pore or the β-barrel, within the membrane. The centre of the pore and the sections of chain protruding from the membrane are likely to have more hydrophilic side groups, since they are in contact with water.

EXERCISE 3.7

The example I thought of was the development of 'electronic noses', which use an array of sensors – for instance, a set of cantilevers with different characteristics – to provide electronic fingerprints for the detection of a range of substances. These can be used for quality control (e.g. in food production) or for detecting dangerous or illegal substances (e.g. hazardous fumes or gas leaks in oil refineries, or 'super-bugs' in hospitals).

EXERCISE 3.8

In Figure 3.24, the sodium and potassium ions are acting as ligands and are shown fitting into binding sites. The phosphate group is also acting as a ligand, although this is less obvious from the diagram.

EXERCISE 3.9

The example I thought of was the development of computer interfaces based on, for example, eye movement, rather than a conventional keyboard. In this example, signals associated with the horizontal and vertical movement of the eye can be picked up by suitably placed electrodes, amplified, and used to control a cursor on the computer screen.

EXERCISE 3.10

The 'TP' in ATP stands for triphosphate, indicating that the molecule carries three phosphate groups. It is when one of these groups breaks off that energy is released and may be used in a reaction. In the process the ATP is converted to ADP.

EXERCISE 3.11

The example I thought of was the development of enzymes that allow raw materials usually obtained from petroleum to be produced from plants. For instance, Sorona® is a new polymer made from propanediol, developed by DuPont as an alternative to polyester and nylon. Propanediol usually comes from petroleum, but an enzyme has been developed to produce it from corn sugar instead.

EXERCISE 3.12

You could have chosen the ATP-synthase rotary motor used to turn a propeller, described at the end of Section 4, or the 'musclebot' described in 'First robot moved by muscle power' at the start of Section 5, which we can assume to be based on myosin. My suggestion for an engineering application is the development of artificial muscle for robotics applications.

EXERCISE 3.13

Assuming that the primer attaches itself to the first two groups so that these do not appear, the sequence that would be obtained corresponds to ATGGACTTGA; see Figure 3.60.

Figure 3.60 The sequence ATGGACTTGA as read by the laser scanner

EXERCISE 3.14

DNA constructions are of a suitable size to provide structural parts for nanoscale robots and could be attached to proteins (such as a molecular motor) by incorporating suitable free ends for interaction. Other examples might be the construction of 2D grids or 3D scaffolding to hold other components in regular arrays (useful for protein crystallography, for example), or the production of templates for tiny electronic circuits.

ANSWERS TO SELF-ASSESSMENT QUESTIONS

SAQ 3.1

(a) Putting everything into SI units and using $t = x^2/D$ gives

(i) $t = (20 \times 10^{-6}\ \text{m})^2/(18 \times 10^{-10}\ \text{m}^2\ \text{s}^{-1}) = 0.22\ \text{s}$ (ii) $t = 0.89\ \text{s}$ (iii) $t = 4\ \text{s}$

(b) My finger measures about 8 cm, which gives a time of 3.6×10^6 s or more than 41 days!

SAQ 3.2

(a) Use the bottom-of-the-range value of the tensile strength to find the minimum diameter, i.e. 1.0 GPa for dragline silk, 0.5 GPa for nylon.

We know that the tensile strength is the force per unit area needed to break the fibre; thus:

$$\sigma = \frac{F}{A} = \frac{F}{\pi r^2}$$

where r is the minimum radius, in m.

Therefore, at breaking point:

$$r = \left(\frac{F}{\pi\sigma}\right)^{\frac{1}{2}}$$

$$= \left(\frac{50 \times 10^{-6}\ \text{kg} \times 9.8\ \text{m s}^{-2}}{\pi \times 10^{9}\ \text{N m}^{-2}}\right)^{\frac{1}{2}}$$

$$= 3.9 \times 10^{-7}\ \text{m}$$

The minimum diameter of the dragline that will support the weight of the spider is 0.78 μm. (You might like to compare this with the picture on the cover of this book.)

(b) Using the same method as in (a), but with the weight equal to (60×9.8) N, gives values for the minimum diameter required to support a 60 kg human of 0.86 mm for dragline silk and 1.2 mm for nylon 6.

(c) We know that:

$$\text{density} = \frac{\text{mass}}{\text{volume}}$$

The volume of a metre of fibre of radius r is equal to $1 \times \pi r^2$.

For dragline silk, using the value for r calculated in part (b):

mass per metre $= 1200\ \text{kg m}^{-3} \times 1 \times \pi(4.3 \times 10^{-4}\text{m})^2$
$= 7.0 \times 10^{-4}\ \text{kg m}^{-1} = 0.70\ \text{g m}^{-1}$

The equivalent value for nylon 6 is $1.4\ \text{g m}^{-1}$, and for steel $4.5\ \text{g m}^{-1}$.

SAQ 3.3

(a) In all the structures the hydrophilic heads of the lipids are in contact with the water, while the hydrophobic tails are not.

(b) The salt would dissolve in the water surrounding the micelle. Petrol is hydrophobic, and would mix with the lipid chains in the centre of the micelle.

SAQ 3.4

(a) For this we need an enclosed compartment within which the crystals can grow. Inverted micelles would provide this, and the size of the cavity can be varied by choosing different lipids. Although liposomes also provide compartments, crystals of other sizes could grow from the surrounding liquid.

(b) Micelles are able to surround the hydrophobic molecules and render them soluble.

(c) The bilayer would be most suitable here; concentration gradients across lipid bilayers are very common in cells.

(d) Here a liposome/vesicle-based delivery system could be used, with the vitamin enclosed within the lipid compartment.

SAQ 3.5

(a) Flagelliform silk and dragline silk.

(b) The presence of helical structures: the elastic β-turn spiral and the 3_{10} helix.

(c) The more elastic flagelliform silk has many more units in the elastic β-turn spiral component, i.e. it contains longer loose springs. Also there are no stiff β-sheet regions in the flagelliform silk.

(d) The β-sheet conveys strength because of the large density of hydrogen bonds between adjacent chains, which keep them fixed in position.

(e) The polypeptide chains are packed closely together in the β-sheet and the presence of different side groups would disrupt this regular arrangement.

(f) There are no elastic β-turn spiral modules in the minor ampullate silk to provide elasticity, but β-sheet regions are present to convey strength.

SAQ 3.6

(a) Random implies no crystallization. An irregular sequence of groups with little capacity for forming secondary bonds is required. The two protein chains must be able to link together at certain points to form a network. The chains must contain some hydrophilic groups to associate with the water molecules and keep them trapped within the network.

(b) Choosing an irregular sequence of groups that won't form secondary bonds is negative design. The linkages, and the need for some hydrophilic groups, are positive design.

(c) If the gel is required to break up on heating then the cross-links must not be permanent, but formed by weak bonds that can break when heated.

SAQ 3.7

(a) The key feature of the primary structure is the sequence of three amino acids that forms the light-emitting centre. The key features of the higher-order structure are the β-can, which protects the light-emitting centre from the polar water molecules that surround the structure and would interfere with its operation, and the α-helix, which positions the light-emitting centre in the right place in the centre of the can (both of these would, of course, require a suitable primary sequence of amino acids to form, but different variations are possible).

(b) The loose loops at the top and bottom of the structure, and the chain ends, would be the obvious places to attempt to modify the structure, as their precise shape is likely to be less crucial to the functioning of the device.

SAQ 3.8

(a) Optically, by reflecting a laser beam from the surface of the cantilever and thus effectively amplifying the signal. Alternatively, electrically using a detector based on a piezoelectric material at the fixed edge of the cantilever.

(b) Rearranging Equation (3.2) gives:

$$f_1 = \left(\frac{4\Delta m \pi^2}{k} + \frac{1}{f_0^{\,2}} \right)^{-\frac{1}{2}}$$

$$= \left(\frac{4 \times 10^{-15}\ \text{kg} \times \pi^2}{0.1\ \text{N m}^{-1}} + \frac{1}{(5 \times 10^6\ \text{Hz})^2} \right)^{-\frac{1}{2}}$$

$$= 1.5\ \text{MHz}$$

SAQ 3.9

(a) Both features mimic the environment of the cell. The planar lipid bilayer provides support for the sensor. Water is needed to ensure that the protein takes up the correct shape to operate. The salt solution provides an ionic environment similar to that in a cell, and also provides the ions that carry the charge when a potential is applied.

(b) To achieve loose binding, the binding site should interact with the target molecule via weak bonds, such as hydrogen bonds or ionic interactions.

(c) Because the bonding between the binding site and the target molecule is weak it is unlikely to be permanent. The molecules will be in constant random motion due to their own thermal motion and collisions from the molecules around them. They are likely to be shaken off before too long. The average period of attachment will be related to the strength of the interaction, but the actual period will vary because of the random nature of Brownian motion.

SAQ 3.10

(a) The signalling molecule must be small, so it doesn't disrupt the membrane too much, and non-polar, so it doesn't stick to the polar lipid heads at the surface and is not repelled by the hydrophobic tails in the centre.

(b) A binding site is a site on the surface of a protein with a well-defined shape and charge distribution, which fits precisely with the shape and charge distribution on a particular ion, molecule or part of a molecule. The ion, molecule or part of a molecule that fits into the binding site is called the ligand. The signalling molecule (or part of it) must be able to act as a ligand, with the right shape and charge distribution to fit into a specific binding site on the surface of the protein.

SAQ 3.11

(a) The time is given by $(10^{-7}\ \text{m})^2/(3 \times 10^{-10}\ \text{m}^2\text{s}^{-1}) = 3.3 \times 10^{-5}$ s or 33 μs.

(b) By calculating the time taken to travel 1 cm in each nerve, I found the difference in time to be 9.9 ms. Dividing this by the answer from part (a) yielded a value for the number of additional synapses per cm in the slow nerve of about 300.

SAQ 3.12

The charge carriers in the axon are positive ions; in a current flowing down a wire they are electrons. The current down a wire is continuous and can have any magnitude; in the axon the charge is passed along in pulses of fixed magnitude, where an increase in intensity is represented by an increase in the number of pulses. In a wire the current flows in the same direction as the potential gradient; in the axon the potential gradient is perpendicular to the membrane along which the charge is propagated.

SAQ 3.13

(a) Energy per ATP/ADP conversion $= 5.1 \times 10^{-20}$ J/$(1.6 \times 10^{-19}) = 0.32$ eV.

This is slightly less than the energy of photons of visible light, which is typically 2–3 eV.

(b) There is a net production of two ATP molecules for each glucose ring split in half. The gain in energy stored in ATP is therefore 1.02×10^{-19} J.

SAQ 3.14

(a) The voltage provided by the battery is 1.13 V. This is the difference in redox potential between the reactions that occur at each end of the electron transport chain. The electric circuit analogy is reasonable in that electrons are travelling along a circuit and providing the power for the pumps to operate, but in the ETC these are specific, single electrons, which is rather different from a conventional circuit, and the ‘wires’ joining the pumps in the ETC are actually mobile molecules that physically shuttle the electrons across.

(b) Six molecules of H_2O require twelve hydrogen atoms, each of which is split into an H^+ ion and an electron at the beginning of the ETC. So, twelve electrons pass along the ETC to produce six molecules of water.

(c) From Figure 3.31, the potential changes across protein 1 from −0.315 V to 0.045 V. The drop in potential is thus 0.360 V and the energy released by each electron is 0.360 eV. Similarly, the energy released across protein 3 is 0.190 eV and across protein 5 it is 0.580 eV.

SAQ 3.15

(a) The axle in the centre of the structure moves, because it is connected to the rotary motor embedded in the membrane. The protein ring around the outside is stationary, because the stator holds it in place. However, the binding sites within this structure undergo slight changes in shape as the asymmetrical axle rotates.

(b) Three molecules of ATP are produced during each complete rotation, so 300 molecules of ATP are produced each second. Since the energy stored in each molecule of ATP is 5.1×10^{-20} J, the power is given by the rate at which energy is converted:

Power = 5.1×10^{-20} J $\times$ 300 s^{-1} = 1.53×10^{-17} W or 15.3 aW (atto = 10^{-18}).

SAQ 3.16

(a) Energy stored in ATP per molecule of glucose = $36 \times 5.1 \times 10^{-20}$ J = 1.84×10^{-18} J.

Efficiency = $((1.84 \times 10^{-18}\ \text{J})/(4.83 \times 10^{-18}\ \text{J})) \times 100 = 38.1\%$.

This is quite a reasonable energy efficiency, much better than is achieved, for example, in a car fuelled by petrol. The 'wasted' energy will mostly end up as heat, so also fulfils a useful function in keeping our bodies at the right temperature for optimum performance.

(b) No. of ATP conversions per day = $10 \times 10^6/(5.1 \times 10^{-20}\ \text{J}) = 1.96 \times 10^{26}$.

Mass of phosphate transferred = $1.96 \times 10^{26} \times 95 \times 1.66 \times 10^{-27}$ kg = 30.9 kg.

30.9 kg is a little under 5 stone, so the statement does seem reasonable.

SAQ 3.17

(a) When ATP synthase works as a motor, ATP molecules attach to the binding sites of the F_1 unit and break down into ADP plus phosphate. The resulting shape changes push on the crankshaft, causing it (and the F_0 unit attached to it) to rotate. The energy is thus transformed from chemical energy (in ATP) to elastic energy (in the shape changes) to rotational kinetic energy (as the crankshaft and the F_0 unit rotate). In the cell, the next transformation would be to electrochemical energy by pumping protons across the membrane. Outside the cell the rotational movement is more likely to be used to do work in moving a load.

(b) The maximum power output possible is the same as you calculated in SAQ 3.15, i.e. 1.53×10^{-17} W or 15.3 aW.

SAQ 3.18

(a) H^+ ion concentration is increased on the inside of the membrane by splitting hydrogen atoms from water into protons and electrons, and by pumping of protons across the membrane by the ETC. The proton gradient is further increased by the removal of H^+ ions on the outside of the membrane, where they combine with electrons from the ETC to form H atoms, which are picked up by NAPD.

(b) It is used for the synthesis of ATP, using ATP synthase.

(c) It transports H atoms to other sites in the cell where they take part in reactions.

(d) Light energy is transformed to electrical potential energy in the electrons passing down the ETC and then to electrochemical energy in the form of a proton gradient. ATP synthase converts the electrochemical energy, via rotational kinetic energy and elastic energy, to chemical energy stored in ATP.

SAQ 3.19

(a) Both collagen and actin are built up from amino acids, or peptide units. Both form long filaments. However, in collagen the polypeptide chains are stretched out and intertwine to form a fibre, as in a rope. In actin the polypeptide chains fold into a more globular shape and the filaments are built up by attaching one unit to the next, more like a chain of beads in a necklace.

(b) The cross-sectional area of an actin filament is given by:

$$2 \times \pi \times (2.5 \times 10^{-9}\ \text{m})^2 = 3.9 \times 10^{-17}\ \text{m}^2$$

The tensile strength is therefore given by:

$$600 \times 10^{-12}\ \text{N}/(3.9 \times 10^{-17}\ \text{m}^2) = 15.4 \times 10^6\ \text{N m}^{-2} = 15.4\ \text{MPa}$$

This is considerably less than the value for collagen, and arises from the relatively weak bonds linking the actin units together.

(c) The globular units in actin are linked together by relatively weak bonds. The filament is most likely to separate at these points when flexed, producing weak points in the structure. In contrast, collagen fibres have strong covalent linkages running along the length of the chain and are largely unaffected by bending.

SAQ 3.20

(a) $W = Fd = (2 \times 10^{-12}\ \text{N}) \times (8 \times 10^{-9}\ \text{m}) = 1.6 \times 10^{-20}\ \text{J}$.

Energy available per ATP/ADP conversion is 5×10^{-20} J.

Efficiency of motor $= ((1.6 \times 10^{-20}\ \text{J})/(5 \times 10^{-20}\ \text{J})) \times 100 = 32\%$.

(b) I have assumed the weight of the apple to be ~1 N. Any value close to this is acceptable. Assume that the force generated per motor is 2 pN, as above.

No. of myosin motors needed $= 1\ \text{N}/(2 \times 10^{-12}\ \text{N}) = 5 \times 10^{11}$.

(c) I did this by bisecting the lower triangle to give two right-angled triangles with base length 12.5 nm. Then x can be found from:

$$\sin 35^\circ = \frac{12.5\ \text{nm}}{x}$$

$$x = 21.8\ \text{nm}$$

This corresponds to a relatively long α-helix with just over 40 turns.

SAQ 3.21

(a) The basic structures of the kinesin and myosin motors are very similar. Each motor is made up of two protein chains twisted together to produce a double-headed structure attached to a long stalk. Both motors rely on shape changes caused by the attachment, break-up and release of ATP, which determine whether or not they bind to a substrate. The substrates to which they attach are different: kinesin interacts with microtubules, myosin with actin filaments (also called microfilaments). Kinesin motors move by 'walking' along the track and are always attached to the microtubule in at least one place. Myosin motors alternately attach and detach, and produce movement by pushing on the actin filaments. Kinesin motors work independently whereas myosin motors work together in large ensembles in muscle tissue.

(b) The force required to detach a myosin head is much less than the tensile strength of actin, so there is no danger of the actin breaking due to the stress placed on it by the myosin.

SAQ 3.22

Power = work done per second = $(7 \times 10^{-12}\ \text{N}) \times (100 \times 8 \times 10^{-9}\ \text{m s}^{-1})$ $= 5.6 \times 10^{-18}$ W or 5.6 aW.

This is about a third of the power available from the ATP-synthase motor.

SAQ 3.23

(a) A codon is a sequence of three DNA bases, which correspond to a particular amino acid. If only two bases were used there would be only sixteen possible combinations, not enough to code for the twenty different amino acids.

(b) 1 $\text{in}^2 = (25.4 \times 10^6)^2\ \text{nm}^2$.

Area per bit $= (25.4 \times 10^6)^2/(100 \times 10^9) = 6.45 \times 10^3\ \text{nm}^2$.

(c) Area of a bit $= (4 \times 0.2)^2\ \text{nm}^2 = 0.64\ \text{nm}^2$

The storage density in DNA is 10000 times higher than in a magnetic hard drive on a computer.

SAQ 3.24

An oscillating cantilever, for example, could be coated with capture probes with the sequence CGAATGCCT. Attachment of a mutated gene would result in a change in the resonant frequency of the cantilever. Other designs using the same capture sequence are also possible.

SAQ 3.25

(a) The tRNA chain is folded up, with only certain portions accessible to interact with other molecules. In the case of mRNA the entire sequence needs to be available to be read, so the mRNA chain is likely to be a linear molecule.

(b) 20: one for each amino acid.

(c) Here RNA is base pairing with another RNA molecule, so the base-pairing rules would be C with G, and A with U.

(d) The base sequence on the mRNA is GGA, which matches a base sequence of CCU on the tRNA. Met is replaced by Gly at the amino acid end.

SAQ 3.26

A ribosome designed to work with DNA would need to incorporate an enzyme for separating the double helix in the region to be read. The DNA codons use T rather than U, so the tRNA molecules would need to pair accordingly.

The main advantage I thought of was that errors occurring during transcription would be avoided, and for a disadvantage I thought that the extra use of the DNA molecule might cause it to be damaged more often; perhaps you thought of others?

SAQ 3.27

There are many different possible answers to this problem. My suggestion is to use six different strands, each twelve units in length, and to construct the loop by designing each chain to overlap with each neighbour over half its length, as shown in Figure 3.61, with the strands labelled A–F.

I decided to use a different six-letter repeat for each half of a chain. For instance, chain A might have the sequence CATGATCTGGCG; half of chain F would then have the sequence GTACTA and half of chain D would have the sequence GACCGC, to pair with chain A. This is positive design. I need to make sure that I don't create more than one opportunity for pairing, so that the arrangement I want is the only one that will occur. This is negative design.

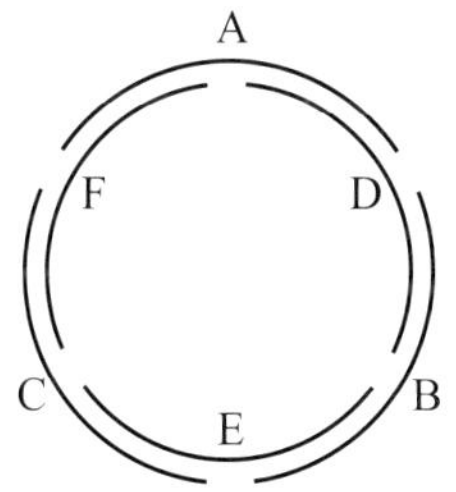

Figure 3.61 Circular loop of DNA double helix

SAQ 3.28

(a) Electron transport chains are found in the membranes of mitochondria and chloroplasts. They use energy from chemical or photochemical reactions to pump protons across the membrane. (Power production.)

(b) Phospholipid molecules arranged in a double layer make up cell membranes, which provide containment for the cell and support for membrane proteins. (Structures.)

(c) Microfilaments are usually found around the edges of cells. They can be used to move cells around and provide tracks for the myosin family of molecular motors. (Structures and transportation.)

(d) Ribosomes are molecular assemblers found inside cells that construct new polypeptide chains from a blueprint provided by mRNA, using amino acids delivered by tRNA. (Manufacture.)

(e) Kinesin motors travel along microtubules within cells, transporting cargo from one place to another. (Transportation.)

(f) Proton gradients are created across the membranes within mitochondria or chloroplasts. They provide a convenient method for storing energy in electrochemical form. (Power production.)

(g) Sodium and potassium ion channels are associated with nerve impulses and found mainly within nerve cells, or neurons. They open and close in response to signals, allowing a pulse of charge called an action potential to pass along the membrane. (Communication systems.)

(h) RNA is a nucleic acid produced in the nucleus of cells from the blueprint carried by DNA. Different forms are produced to carry the instructions for making a protein (mRNA), to deliver amino acids to the ribosome that assembles the new polypeptide chains (tRNA) and to form part of the ribosome itself (rRNA). (Manufacture.)

(i) ATP synthase is an enzyme that operates as a rotary motor and generator to convert electrochemical energy stored in a proton gradient to chemical energy stored in ATP. It can also work in reverse to generate rotary movement using chemical energy from ATP. (Power production.)

(j) Axons are extended protrusions from nerve cells (neurons) that carry the electrical signals that form the basis of the nervous system. (Communication systems.)

SAQ 3.29

Here I will suggest just one example from the text. In most cases there are several possible alternatives.

(a) Green fluorescent protein has a primary structure resulting from the sequence of amino acids along the chain, and a higher-order structure in which the polypeptide is folded up to form a can, containing portions of α-helix and β-sheet.

(b) The flow of electrons along an electron transport chain is an example of a flow of electric current within a cell.

(c) (i) Chemical energy is converted to mechanical work when kinesin motors transport cargo along a microtubule.

(ii) Electrochemical energy from a proton gradient is converted to chemical energy in ATP by ATP synthase.

(iii) Light energy is converted to electrochemical energy in the form of a proton gradient in the chloroplasts within plant cells.

(d) Rotational motion is provided by ATP synthase used to turn a metal propeller.

(e) Kinesin motors attached to a glass slide can move microtubules around.

(f) Enzymes can break the chain, join two portions of chain and use one strand of a chain to synthesize another.

(g) DNA can be used to create physical structures such as lattices and cubes.

(h) Virtually every protein we have discussed could be regarded as acting as a catalyst, and described as an enzyme. The term is most often applied to chemical reactions where one substance is broken down, so the digestive enzyme pepsin would be a good example.

Acknowledgements

Grateful acknowledgement is made to the following sources:

Text

Pages 59–61: Ananthaswamy, A. (2004) 'First robot moved by muscle power', *New Scientist*, vol. 181.

Figures

Figure 3.15: From Isaac Leung's page, www.physics.ubc.ca/~isaacl/, The University of British Columbia.

Figure 3.16: Structure of Green Florescent Protein. Taken from: www-bio.rice.edu. Department of Biochemistry and Cell Biology, Rice University, Houston.

Figures 3.17 and 3.20: Adapted from Bayley, H. and Cremer, Paul S. (2001) 'Stochastic sensors inspired by biology', *Nature*, vol. 413, Macmillan Magazines Ltd.

Figures 3.29 and 3.37: Courtesy of Professor Mike Stewart, Department of Biological Sciences, The Open University.

Figures 3.33 and 3.51: The Royal Swedish Academy of Sciences.

Figures 3.34, 3.35, 3.57 and 3.58: Adapted from Goodsell D. S. (2004) *Bionanotechnology – Lessons from Nature*, Copyright © 2004 Wiley-Liss, Inc. This material is used by permission of John Wiley & Sons, Inc.

Figure 3.41: Courtesy of The Nanostructures Group, Bristol University.

Figure 3.44 and 3.45: Valel R. D. & Milligan R. A. 'The Way Things Move: Looking Under the Hood of Molecular Motor Proteins', Department of Mechanical Engineering, Massachusetts Institute of Technology.

Figure 3.48 (a): © Anne Arlidge and Jancis Smithells; (b): © Charles C. Benton; (c): © Abigail Fallis/Cass Sculpture Foundation.

Figure 3.53: 'DNA microarrays – the diagnostics of the future', www.ogham.de.

Course team acknowledgements

Block 3 *Working with nature the cell as an engineering system* was prepared for the course team by Dr Sally Organ.

T356 course team

ACADEMIC STAFF

Dr Alec Goodyear (course chair)

Professor Nicholas Braithwaite

Jan Kowal

Dr Tony Nixon

Dr Sally Organ

Robin Harding (critical reader)

James McLannahan (critical reader)

Dr Martin Rist (critical reader)

Dr George Weidmann (critical reader)

Peta Jellis (course manager)

EXTERNAL ASSESSOR

Professor John Fothergill, Professor of Engineering and Pro Vice Chancellor (Learning and Teaching), University of Leicester

SUPPORT STAFF

Debbie Derbyshire

Colin Gagg

Gordon Imlach

PRODUCTION STAFF

Annette Booz

Philippa Broadbent

Lisa Carrick

Daphne Cross

Sue Dobson

Henry Dougherty

Anna Edgley-Smith

Charles Harris

Peter Heatherington

Carol Houghton

Jonathan Martyn

Katie Meade

Lara Mynors

Jonathan Owen

Deana Plummer

Lynn Short